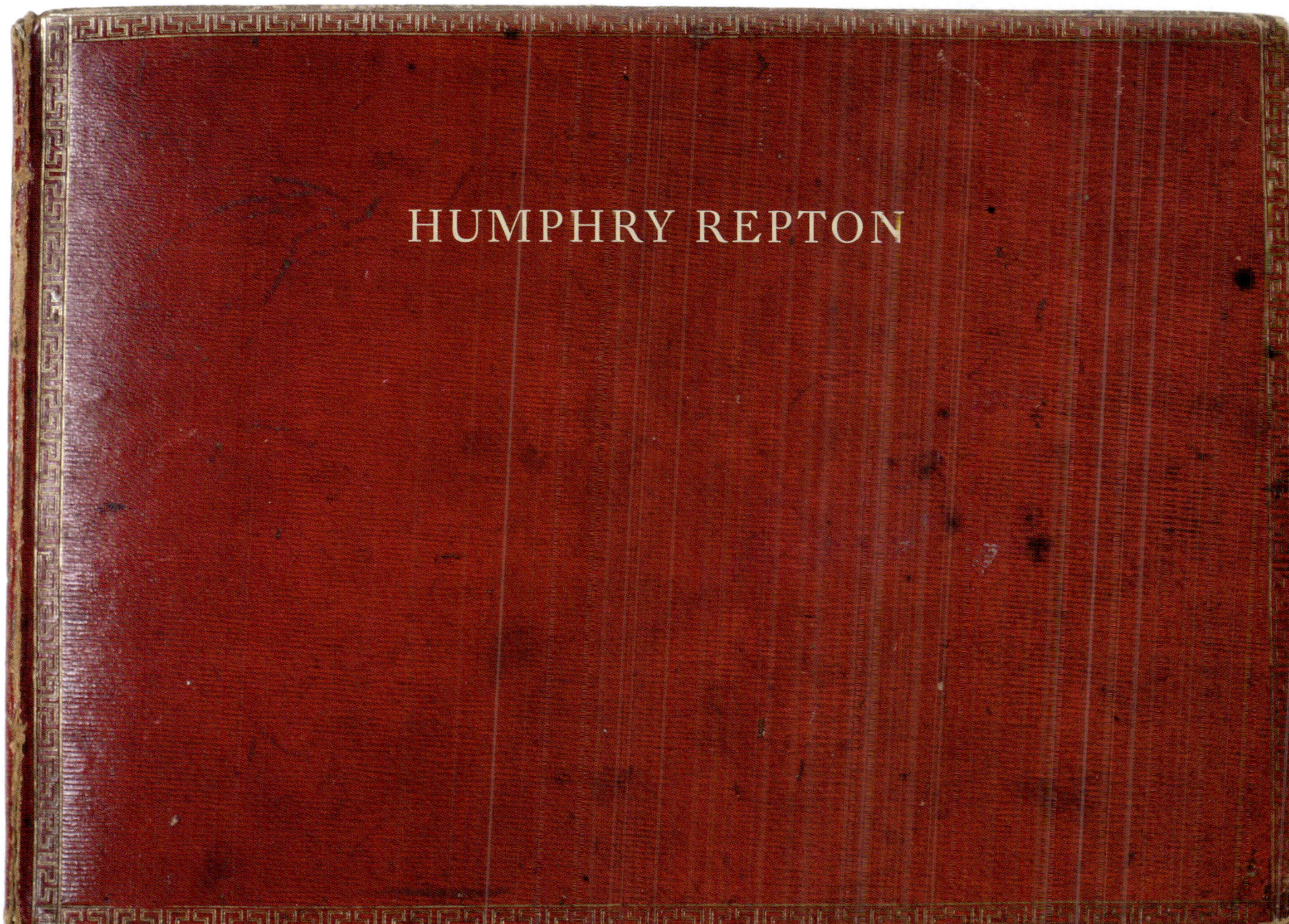
HUMPHRY REPTON

JOHN PHIBBS
PHOTOGRAPHY BY JOE CORNISH

HUMPHRY REPTON

Designing the Landscape Garden

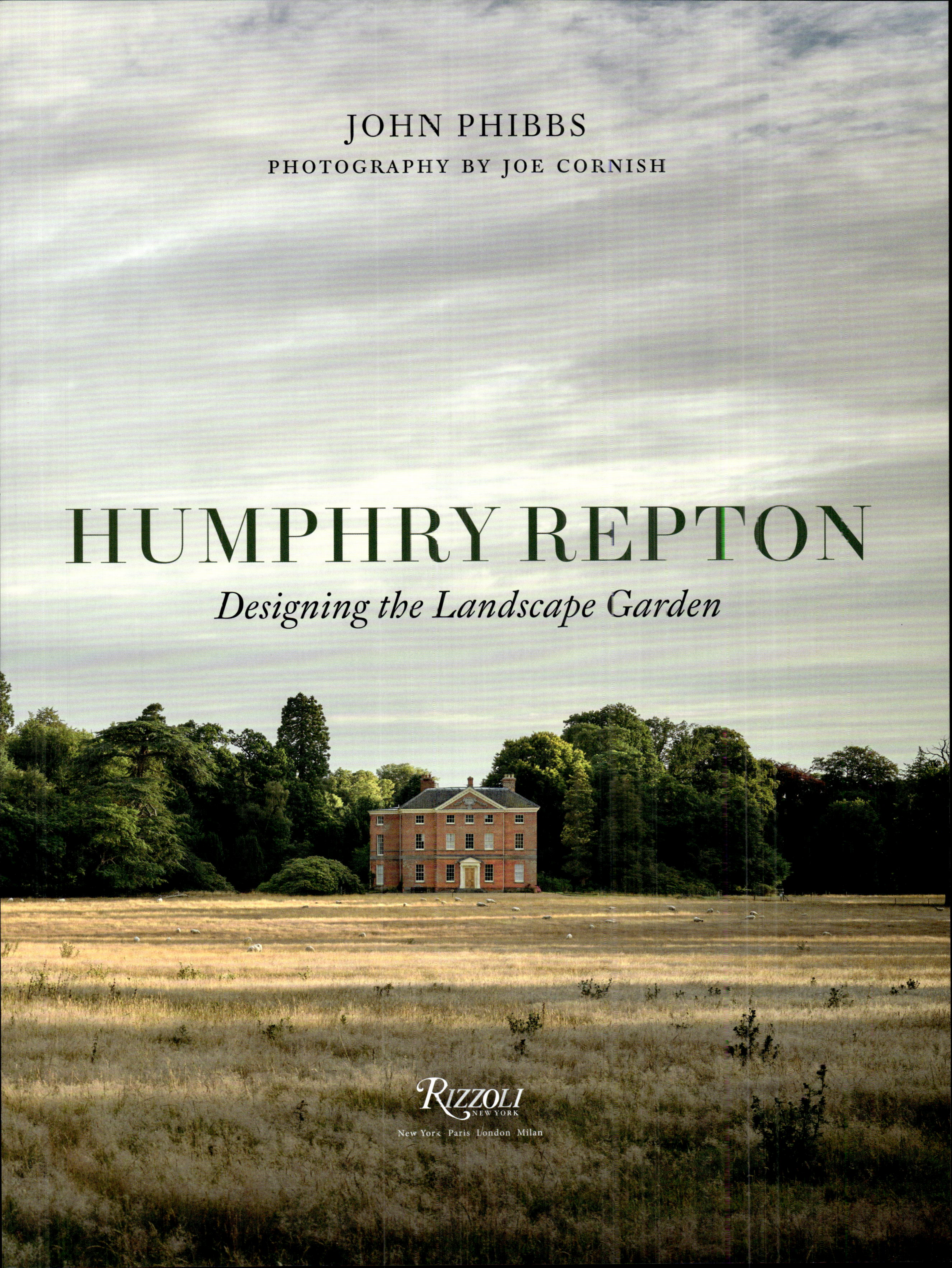

RIZZOLI
NEW YORK
New York Paris London Milan

First published in the United States of America in 2021 by
Rizzoli International Publications, Inc.
300 Park Avenue South, New York, NY 10010
www.rizzoliusa.com

Publisher: Charles Miers
Editor: Jacob Lehman
Production Manager: Colin Hough-Trapp
Production Assistant: Olivia Russin
Managing Editor: Lynn Scrabis
Copy Editor: Gail Spilsbury

All photographs by Joe Cornish
with the exception of images on the following pages:
18, © National Portrait Gallery, London; 32–33, 36, 37, 39 courtesy the Morgan Library and Museum, New York (gift of Mr. Junius S. Morgan and Mr. Henry S. Morgan); 59, 62 © the Trustees of the Wormsley Fund; 99, 102, 103 © Historic England Archive; 170, 171, 172 courtesy Bristol Culture, Blaise Museum; 269, 270–271, 274, 279 from the Woburn Abbey Collection, © Bedford Estates.

Designed and typeset in Fleischman by Dalrymple

Printed in China

2021 2022 2023 2024 / 10 9 8 7 6 5 4 3 2 1

ISBN: 978-0-8478-6354-9
Library of Congress Control Number: 2021935670

Facebook.com/RizzoliNewYork
Twitter: @Rizzoli_Books
Instagram.com/RizzoliBooks
Pinterest.com/RizzoliBooks
Youtube.com/user/RizzoliNY
Issuu.com/Rizzoli

Opening pages: a visual resumé of Repton's career.

Pages 2–3, 4–5: In his red book for Babworth, Nottinghamshire (1790), Repton sought to make an appropriate first view, or burst, to the house.

Pages 6–7: At Honing Hall, Norfolk (red book 1792), the land is flat and the house modest. Repton kept the park small.

Pages 8–9: The Dry Arch at Mulgrave (red book 1793). The structure was designed to match the character of the ruins of the ancient castle.

Pages 10–11: At Rûg (red book 1795), Repton made a large lake in the Brownian style deep in the heartland of Capability Brown's greatest critics.

Pages 12–13: The Limes at Sarsden (red book 1796). Repton liked to plant small groups of trees at 10'–20' intervals.

Pages 14–15: Late in his career, at Endsleigh (red book 1814), Repton made his most perfect, joyful designs; here the Children's Garden, with a late summer planting of salvias.

S. Shelley. del[t] H.B.Hall. sc[t]

HUMPHRY REPTON, ESQ.

THE UNLIKELY GENIUS OF HUMPHRY REPTON

Some are born to greatness, some achieve greatness,
and some have greatness thrust upon them.

WILLIAM SHAKESPEARE

Picture yourself travelling through England on a train, through woods and fields, past hills and villages, until the ribbons of suburban development tell you that you are nearing your destination. At that point, as you are about to put on your coat and pull your suitcase from the rack, the man sitting opposite you leans forward and taps your knee to get your attention. He says that all these rows of semidetached houses, with their little front gardens onto the street, and in back, the kids' swing in a single apple tree, the row of vegetables, and the garden shed, that all this England was conceived by one man, Humphry Repton; and that his understanding of the way each social class should live has drawn a blueprint for the entire English-speaking world—for Canada, for Australia and New Zealand, for South Africa, India, and the United States of America.

If your response to such an approach would be to pull back from the mad fantasist, grab your case, and leave the train immediately, then this book is for you. For Repton (1752–1818) is the last and the most influential of three giants—the other two being William Kent (1685–1748) and Lancelot "Capability" Brown (1716–1783)—and now, 200 years after his death, is the time to celebrate his unique achievement.

These three men formed and led the English Landscape Movement in the eighteenth century, a style that still has a global impact on our relationship with the natural world, but Repton came late to landscape and more or less by accident. In June 1788, he was thirty-six, with little money, a wife, Mary, and six children, living hugger-mugger in a cottage on a main road in Essex. Then he had a dream:

> *The possibility of turning to advantage that natural taste for improving the beauties of scenery, which had formed one of the dearest pleasures of his rural life, suggested itself to his mind one night when anxiety had driven sleep from his pillow. This scheme, which at first seems to have entered his imagination with almost the vague uncertainty of a dream, assumed a more substantial form when, with the return of day, he meditated upon its practicability. With his usual quickness of decision, he arose the next morning; and, with*

opposite: Samuel Shelley painted Humphry Repton in his prime, in about 1802; the image was later engraved and was published in the 1840 edition of Repton's collected works on landscape.

fresh energy of purpose, spent the whole of that day in writing letters to his various acquaintances, in all parts of the kingdom, explaining his intention of becoming a "Landscape Gardener"; and he lost not a moment in bending his whole mind to the acquisition of such technical knowledge as was necessary for the practical purposes of such a profession.[1]

Repton had no professional training in landscape design, but he was a sociable and amusing companion who shared the values of his clients—he could read French and a smattering of Latin, he could paint and play the flute, and he was comfortable discussing the latest scientific theory. He was always bustling about, usually cheerful, cocksure, "a gentleman ... of so many words that he is not easily shaken off; he asserts so much, and assumes so much, as to make me irritable, for he is one (of the many) who is never wrong." Above all, he had taste and he had contacts, which guaranteed him immediate success. As immediately, however, he found himself enmeshed in controversy, and his memoir and later letters suggest an old man depressed about the state of the nation, about his own work, and about his legacy. These writings encourage a reading of him as somehow second-rate.

The magic of what might otherwise be a depressing biography, however, is that with every setback, Repton's work became greater, more complex, and more profoundly charged. Under the façade that he presented to the world, he wrestled throughout his career with two major questions: what kind of architecture is best suited to country houses? and how could ordinary countryside be made the setting for those houses (rather than that idealized form of nature that Brown pursued for much of his career)? Those magnificent landscapes that survive from his final years—Endsleigh, Ashridge, and Woburn Abbey among them—are his answer. They compare with the greatest of Capability Brown's achievements and, unlike Brown, the impact of what he thought about landscape was immediate. The model of housing for working people that he promoted was adopted throughout the English-speaking world, and he was venerated by the great American landscape architect, Frederick Law Olmsted, who asked his protégé, Charles Eliot, to photograph Repton's cottage when he visited England in 1886. An edition of Repton's writing was the first volume published by the American Society of Landscape Architects in 1907 and this Society was first known as the "Repton Club."

Reduced to a few sentences in this way, his career may read like a picaresque, those novels of moral sentiment, at which Maria Edgeworth was so expert, in which a likeable hero or heroine, good-hearted but naive, sets out to explore the world and through a series of misfortunes, usually caused by his or her own errors, comes to some degree of wisdom and good sense.

In so far as Repton's story resembles a novel, it requires a historical context and involves a cast of characters. Certain things also need an introduction:

Each of Repton's RED BOOKS relates to a single landscape commission and consists of a handsomely bound collection of sketches, plans, and text in which Repton set out his proposals. They are now greatly valued as works of art in their own right, but they also make an invaluable contribution to our understanding of his landscapes.

A SUNK FENCE is a field boundary that is hidden from sight. Sometimes the fence or hedge is run along a deep ditch; sometimes there is a "ha-ha," that is a ditch with a retaining wall on one side.

As for context, two events and a theory stand out:

THE GLORIOUS REVOLUTION (1688–9) took place 100 years before our story starts and involved the "bloodless" overthrow of the Catholic James II by his Protestant daughter Mary and her husband, William of Orange. William's supporters were known as "Whigs." "Tories" originally supported James II and his heirs.

THE BILL OF RIGHTS (1689) created the separation of powers in the English constitution, limiting the powers of monarchy and enhancing those of the

opposite: The banks at Mulgrave, through which Repton brought his south approach, are cut through with picturesque streams and waterfalls.

democratically elected House of Commons and of the aristocracy in the House of Lords. The separation of powers remains a cornerstone of British and American democracy, indeed of every democracy.

THE ENGLISH rapidly came to value and take pride in the liberty that these two events provided.

THE PICTURESQUE MOVEMENT was a theory, influential in the late eighteenth and early nineteenth century, that landscape gardeners should model their work on landscape painting. It generated a good deal of discussion, which was known as the Picturesque Controversy.

As for characters, a few are listed here who played their part in Repton's life:

JANE AUSTEN (1775–1817) discussed landscape gardening in her novels *Sense and Sensibility* and *Northanger Abbey*, saw Repton's work at Stoneleigh Abbey, and mentioned him by name in *Mansfield Park*.

NAPOLEON BONAPARTE (1769–1821) requires no introduction.

LANCELOT "CAPABILITY" BROWN (1716–1783) was the titanic genius of English landscape, against whom, in Repton's time, there was to be a backlash.

EDMUND BURKE (1730–1797), statesman and political theorist, author of *A Philosophical Enquiry into the Origin of Our Ideas of the Sublime and Beautiful* (1756), as well as *Reflections on the French Revolution* (1790), was an acquaintance of Repton and eventually a close ally of William Pitt the Younger.

SIR WILLIAM CHAMBERS (1723–1796) was an architect, theorist, critic of Capability Brown, and herald of the Picturesque style.

above: Repton stuck his trade card into his red books. He portrayed himself as a practical man, overseeing the excavation of a deep bay for a lake in the style of Capability Brown.

The Reverend William Gilpin (1724–1804) was the father of the Picturesque with William Mason, not to be confused with his nephew and devotee of the Picturesque school William Sawrey Gilpin (1761–1843).

Samuel Lapidge (1740–*c.* 1806), Capability Brown's leading foreman and subsequently the executor of his will, was a rival Repton had to defeat. Lapidge's career faltered in the 1790s, and he took to drink.

The Rev. William Mason (1724–1797), father of the Picturesque with William Gilpin, wrote the poem "The English Garden" often quoted by Repton, who visited Mason and sought his support.

Robert Marsham (1708–1797), Squire of Stratton Strawless, Norfolk, was a lover of trees and a friend of Repton's. He suggested that Repton paint "before and after" pictures of the landscapes where he worked, a technique that Repton deployed with great success in his red books.

John Nash (1752–1835), architect, enters our story midway through as hero and rapidly leaves it as villain.

William Pitt the Younger (1759–1806), prime minister, enters our story as hero, charismatic and always optimistic; he turns bad then dies, but saves the country. He was the Churchillian son of the Churchillian statesman, the Earl of Chatham, who had led Britain to victory over France in the Seven Years War; even his rival Charles James Fox acknowledged his brilliance.

Sir Uvedale Price (1747–1829) and Richard Payne Knight (1751–1824), protagonists in the Picturesque Controversy and Herefordshire Squires, they begin as the evil villains but come to be seen, by the end, as not so evil after all.

Humphry Repton (1752–1818), himself—an unlikely genius.

Mary Repton (1749–1827), devoted wife of the above, and mother of seven children who survived to adulthood.

John Adey Repton (1775–1860), eldest son of Humphry and Mary, architect, antiquarian, and collaborator with his father.

George Stanley Repton (1786–1858), younger son of Humphry and Mary, architect and cottage specialist, worked for John Nash.

Jean Jacques Rousseau (1712–1788), one of the most original and fecund thinkers of the eighteenth century.

William Windham (Felbrigg Hall), the Duke of Portland (Welbeck Abbey and Bulstrode), leading politicians, supporters of Pitt, hostile to France, and well-wishers of Humphry.

Finally, Repton published six books on landscape that were collected together into a single volume by John Claudius Loudon: *The landscape gardening and landscape architecture of the late Humphry Repton, esq. Being his entire works on these subjects* (London: Longman & Co, 1840). All quotations from his published works are taken from this volume. These works were:

Sketches – *Sketches and Hints on Landscape Gardening* (1795)

Observations – *Observations on the Theory and Practice of Landscape Gardening* (1803)

Enquiry – *An Enquiry into the Changes in Taste in Landscape Gardening* (1806)

Pavilion – *Designs for the Pavilion at Brighton* (1808)

Fragments – *Fragments on the Theory and Practice of Landscape Gardening* (1816)

Quotations from letters and manuscripts (including Repton's red books) retain the case, spelling, and punctuation of the originals.

1789–1792

LEARNING ON THE JOB

In 1788, the London Revolutionary Society, a congenial group of Whig democrats, celebrated the centenary of the Glorious Revolution and the accession of William of Orange, which had brought a new era of harmony, peace, and stability to Britain.

In France, a year later, on June 17, 1789, another revolution broke out. The Bastille was stormed on July 14 and the London Revolutionary Society sent their congratulations. The press hailed the event as a continuation of the Glorious Revolution and the Bill of Rights,[1] and in February 1790, William Pitt the Younger, the prime minister, expressed a general sentiment in the House of Commons: "The present convulsions of France, must, sooner or later, terminate in general harmony ... thus circumstanced France would stand forward as one of the most brilliant Powers in Europe; she would enjoy just that kind of liberty which I venerate."

Two years later, on February 17, 1792, Pitt could announce in Parliament that "there never was a time in the history of this country, when, from the situation of Europe, we might more reasonably expect fifteen years of peace, than we may at the present moment."[2] Edmund Burke, with astonishing prescience, published his account of the dangers of the French Revolution at the end of 1790,[3] but the intoxicating joy of these precious years, 1789–1792, was best captured by William Wordsworth when he recalled them a decade later: "Bliss was it in that dawn to be alive, / But to be young was very heaven!"[4]

Repton could not have chosen a more auspicious moment to launch his scheme: Capability Brown's death in 1783 had left a gap that none of his foremen could fill, the country was brimming with optimism, and through the good offices of his friend and landlord, William Windham, Repton had met and befriended leading political figures, such as the Duke of Portland. Repton's career took off like a rocket. By May 1792, in his red book for Honing Hall, he claimed that he had already written a hundred other such red books. It must have seemed easy work, requiring only the sale of good taste to those who lacked it—or lacked the time to devote to its display—knocking out landscapes in the familiar Brown tradition and passing time with the great and famous. All he had to do was work out what he believed to be Brown's formula, and by 1792, he felt he had done that.

opposite: Across the dingle to the south at Ferney, there was a dramatic stone quarry. This was to be Repton's "enchanted ground."

FERNEY HALL

— SHROPSHIRE —

CHAPTER ONE

FERNEY HALL

Repton's first visit September 26, 1789,
red book October of the same year

FERNEY HALL LIES in a quiet corner of Shropshire today, but in Repton's time and for one hundred years before him, Shropshire and the Welsh borders had seethed with the entrepreneurs of the Industrial Revolution. The world's first iron bridge had been built (at Ironbridge) only twenty-five miles away in 1775. The Ferney estate had been developed by the Walker family in the seventeenth century. They owned the great ironworks and foundry at Bringewood, and they also undoubtedly developed the numerous stone pits to be found around Ferney. This was a landscape scarred in the search for profit.

Repton's client, however, was Samuel Peckham Phipps, a London barrister "of the highest character," according to Repton, who had bought the property two years earlier in 1787. His Ferney was unpromising and small, with under 70 acres available for improvement, most of which was common land, available for some tree planting, but little else. Furthermore, a public road ran from east to west, within fifty yards of the house, cutting it off from Shelderton Common to the north. Its ornamentation, when Repton arrived, was limited to modestly terraced gardens on the east front, some garden walls, and a series of terraced walks that hung down the steep slope on the south side of the house. But there were less contrived assets: a 22-acre orchard of Spanish chestnuts on the relatively level ground of Shelderton Common to the north and west (presumably the Walkers had been hoping for a profit from these trees); while to the east, the ground sloped gently to a distant view of the town of Ludlow five miles away across Duxmoor Common. Immediately south of the house, below the terraced walks, was a dingle, or deep valley, with a pond and stream on its floor; on the rising ground beyond were the quarries in Stonepits Coppice; above them, the open hill of Mocktree Common; and beyond, the great neighbouring estate of Downton Castle. This had been assembled by another industrialist, Richard Knight (1659–1745), who took over the forge at Bringewood, modernized it, and made himself a fortune. Knight must be mentioned because his grandson was the connoisseur Richard Payne Knight. Repton published an account of meeting this younger Knight and his friend Sir Uvedale Price; the three men explored the Wye amicably together, but fell out at Ferney, and Price and Knight, who went on to found the Picturesque Movement, were to become Repton's increasingly bitter critics.

The work Repton proposed here looks unexceptionable. In essence, he was reproducing ideas that he might have found in one of Capability Brown's

previous pages: The boxed shrubs are recent, but the trees below Duxmoor Common crowd out Ludlow and turn the view to the left, to the land and distant hills beyond.
opposite: Ferney's seventeenth-century chestnut orchard stands unannounced beside the Ludlow road.

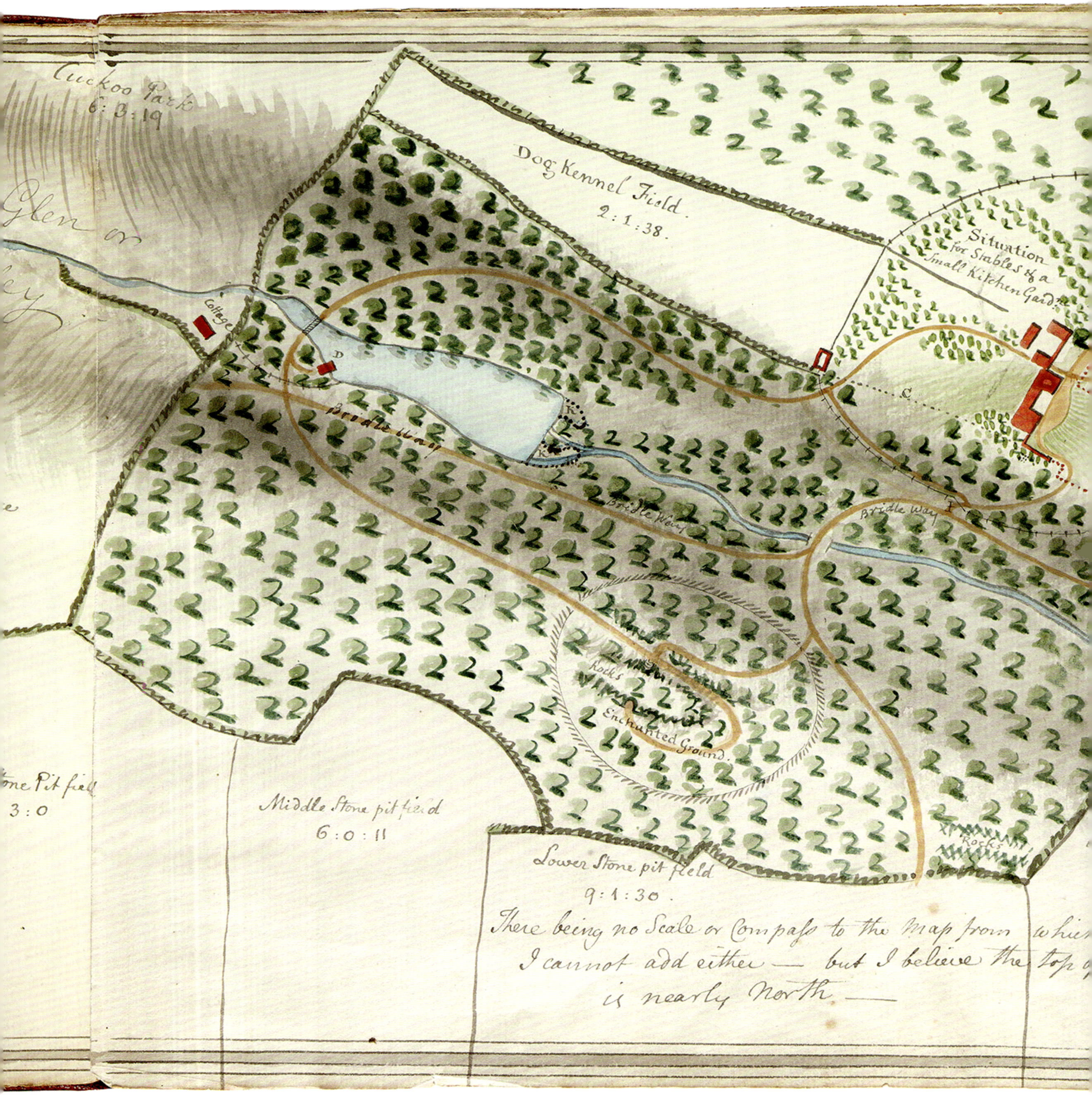

previous pages: The house sits immediately below the chestnut orchard on a shelf cut into the side of the dingle.
above: Repton sought to fence the "enchanted ground" of the quarry, and the system of drives indicated by the orange lines on his plan was to be further elaborated with loops running west above the lodge.

landscapes from the 1760s: first, there were the old, formal terraces on the east and south fronts; these had to go so that Repton could restore "the natural form of the ground." Then, he had to tidy up the buildings around the house, find space for a new stables and new kitchen garden, and put in a sunk fence to open out the view east; after that, an extension to the pond in Ferney Dingle to something more like a lake; then, though very vague in his instruction, to lay out some walks through the valley and to plant a few trees on Duxmoor Common. He would have liked to add to the domain the woods and Mocktree Common on the high ground south of the dingle, and to build a shepherd's lodge or tower on the common directly opposite the house; also to move the approach from Ludlow, and—his pièce de résistance, though again without any detail given—to turn the biggest of the quarries ("the magic chasm" with its "divided rocks which project like the shatterd [*sic*] walls of some ruin'd Castle") into an "Enchanted Ground."

In all the red books that he wrote in the eighteenth century, Repton assumed that he and his clients were of one mind, that in "former Times" there had been "false taste," but that he and his contemporaries shared the same opinion of what taste was and what landscape was, including—at least in 1789—that Brown's work was the paradigm, with its extensive grassed parkland smooth enough to mow and its large bodies of placid water. These were the two elements for which Knight and Price reserved their strongest criticism, dismissing them in their hostile publications of 1794 as "monotonous." Indeed, Price singled out for criticism the "rocky dell" where Repton proposed to extend the lake and add "a small bath house" as one of the elements that dominated the argument that Knight had with Repton at Ferney in 1789.

Repton's struggle with their theory of the Picturesque coloured his work for the next twelve years or more, steering him away from the creation of parkland and large placid lakes. We have to imagine the details of their conversation, but Repton's commission at Ferney also introduces the danger of relying solely

overleaf: Repton's lake can never have been satisfactory. The ground falls steeply and a very high dam would have been needed to give the water a reasonable surface area.

on his red books as an account of his achievement. Thus his plan showed the road that between the house and the Sweet Chestnut grove on its north side as diverted. This did happen but not until 1822, four years after his death, and we can be sure that the works that he showed as complete on his plan could not even have been begun in 1789. In short, a red book cannot be regarded as a reliable guide to the situation before Repton started work. Ferney's plan was crudely and rapidly sketched, and it seems likely that Repton added it to the red book in anticipation of the road closure.

Repton's red books carry much advice that is devoted to the convenience and comfort of his clients, to drying-yards and the back road to the stables. In consequence, it is easy to overlook his success at

above: Repton's proposals for the quarry included a rickety bridge across a chasm some fifteen feet deep.
opposite: By removing the walls and farm buildings, Repton gave the house a setting in the style of his predecessor, Capability Brown.

making landscapes as expressive as his predecessor Capability Brown's, and it is always as well to look over the landscapes themselves for evidence of his hand. So if we read the red book as a record of all that Repton proposed, we would conclude that he did very little at Ferney. Since his client, Samuel Phipps, died in 1790, that might seem a likely conclusion, particularly as a succession of early deaths meant that there was little chance of any major investment until the old house was pulled down and replaced by W. Hurt Sitwell in 1858.

However, Phipps obviously approved Repton's work—he expressed himself "highly satisfied" with the original red book and furthermore, in 1792, his cousin, Francis Sitwell, gave £5,000 to his two cousins: Thomas Peckham Phipps at Little Green and Sir Charles Wake at Courteenhall. Repton was commissioned to work at both in 1793, as though prompted to do so by these gifts. Funds also seem to have been set aside for completing work at Ferney, and Repton's account was not settled until January 28, 1793.

Finally, we have the evidence of Bertie Greatheed, on tour from his own estate in Warwickshire. He was dejected and did not enjoy his visit to Ferney in 1806, claiming to see "nothing but common with a most ugly whitewashed house within a brick walled garden. the Clee hills and Ludlow in front were fine but the weather was thick." The view to Ludlow was Repton's; the brick-walled garden had been reshaped and the whitewash proposed by him: "I have represented the house of a Stone Colour, because red Bricks are utterly incompatible with all ideas of picturesque beauty, and the changing the Colour of a house from red to white, by the Stucco wash is so cheap, & so durable." Above all, the disgruntled Greatheed reported the advice that "If I had gone into the dingle, I should have seen the finest view in Shropshire."[1]

In fact, the two best surviving elements of Repton's design for Ferney barely get a mention in his red book: first, his treatment of Duxmoor Common—"I hope there will be no difficulty in obtaining leave to plant a few trees (fenced with Cradles) on the swelling

above: Ludlow in the shadow of the Shropshire hills. Repton concluded in 1800 that "all natural objects such as Wood, Water, Lawn, and distant Hills or Mountains, appear best with the sun behind them."

knowl of the Common," and second, his plan for the walks and drives in the dingle—"... other walks may be conducted thro' the wood to such objects of 'rare wonderment' as it would be unpardonable not to connect with the house by easy access."

These walks and drives writhe around the dingle and reach as far west as Shelderton at its head. Some were mapped by the Ordnance Survey and others partly survive as earthworks. They would have been exciting to drive around, and in addition to showing off the best views, they make the dingle dynamic, bring it to life, and force the visitor to engage with it. They are thoroughly Reptonian in character; there are drives like them at Mulgrave, Stanage, Blaise, and Endsleigh. Brown's drives are often straighter and can be steeper because carriages in his day tended to be heavier and hence needed more than one horse.

The view to Duxmoor Common, however, is masterly. Repton recommended replacing the central section of the old garden wall with a sunk fence. The final result is not as shown in the red book, but stub ends of the wall, decorated with urns ("I have also suppos'd this wall terminated by a pedestal & Vase") made a proscenium. With the ancient yews of the Dark Walk on the south side, and a shrubbery dominated by a copper beech (which only fell a few years ago) against the wall of the Fruit Garden to the north, the effect was chiaroscuro—the view running out of the dark and onto the bright sunlit grassy stage of the common, where Repton's clients could do little because the land was outside their control. Here, however, Repton created a landscape by the addition of a single horse chestnut. This tree has three functions: first, it conceals the town of Ludlow; second, it leaves on show (from the main window of the house) the great tower of St Laurence, Ludlow's mediaeval parish church, as an object off to the right of the composition; and third, it leads the eye away to the left, away from Ludlow, by gently reinforcing the curve of the slight valley that runs from the house across the common to the northeast. Four additional trees—two oaks, an ash, and a sycamore—on the far side of the common indicate the end of the landscaping without cutting off the open countryside beyond. The restraint is masterly. Though it is simple and unsophisticated, this does not read as the work of an amateur. It is also conspicuously theatrical—not a trait one could often attach to a landscape designed by Capability Brown—an early result, one might think, of his conversations with Price and Knight.

above: Repton broke up the panoramic view from the house with single trees so as to show off the striking hills and the church tower.

BABWORTH HALL

— NOTTINGHAMSHIRE —

CHAPTER TWO

BABWORTH HALL

Repton's first visit April 9, 1790,
red book fifteen days later

At Babworth, Repton found a place that was unimproved but for a few young plantations of trees. However, it did have a delightful situation, ideally suited to the Brownian style. The hall overlooks a beautifully shaped valley to the east, with water at the base, a steeper, wooded slope beyond, rolling parkland to the north and west, and an estate village to the south. Repton made return trips to Babworth and added to his red book when he did. He befriended his clients, the Simpsons, and he was obviously satisfied with what he had accomplished there, yet he evidently rejected this opportunity to show them his mastery of Brownian design.

It is often suggested that Repton was a second-rate designer, offering second-rate solutions to second-rate families. At Babworth, he even classed the hall as second-rate, describing his hopes for it as the "residence of comfort, hospitality & elegance, a situation in the scale of affluence immediately below the cumbrous magnificence of irksome grandeur." Hence, he was concerned not to do too much because "every alteration which may partake too much of splendour on one side, or too little of elegance and convenience on the other will alike offend as being out of harmony with the whole." However his client, the twenty-six-year-old John Simpson, was well set in "the scale of affluence." He was the third son of Sir Henry Bridgeman and had been brought up at Weston Park, one of the great houses of Staffordshire. Simpson was his mother's name, and he had adopted it when he succeeded his uncle, Lindley Simpson, at Babworth in May 1785. In addition to that, John Simpson's wife, Henrietta, was the daughter of Sir Thomas Worsley, who had his own estates in Yorkshire and at Appuldurcombe on the Isle of Wight. John had already been a Member of Parliament, was shortly to become High Sheriff of Nottinghamshire, and was well-known and respected in the county as an agricultural improver on land that had only recently been enclosed from the forest. Henrietta unfortunately died in 1791, but John's second wife, Grace Estwicke, was the daughter of the equally wealthy plantation owner Samuel Estwicke. In short, John had the connections and the money. Furthermore, Weston Park and Appuldurcombe were two of Capability Brown's most magnificent and extensive landscapes.

Another landscape gardener (unnamed, but referred to by Repton) had already drawn up a proposal to make a Brownian park at Babworth, and Repton may have wanted to persuade Simpson that splendour was not as valuable as convenience in order to distance himself from his rival's efforts.

previous pages: Babworth House had been a muddle of old building that Repton sought to make regular.
opposite: Repton's watercolours show a crowded scene that was to be simplified with the removal of the arable land.

This is how we should read his criticism of "mere gardeners or engineers ... who had never studied the sublime requisites of composition in landscape, and who were incapable of tracing the effect of beautiful and pleasing objects to their original causes."

Great landscaping required good taste, and Repton still regarded taste as a universal attribute "preexisting in the nature of the human mind." His critics, Price and Knight, would have added that it could only be cultivated by intelligent and educated people through the study of great art and "the sublime requisites of composition in landscape."

Repton's unnamed predecessor had set out the line for "a most magnificent lake" in the valley. So Repton condemned this (in verse that he later published), concluding that: "There is nothing in which modern improvers have so much erred as in supposing that because a wide expanse of water is in some situations desirable, it must always be so."

A proposed new grass approach fared no better: "I think that a grass road of thirty feet wide, however beautiful it may be over the vast lawns of Clumber—when such an one crosses a field of 13 acres, it would be as incongruous as would be the elegant bridge of that place thrown over the little brook at Babworth," even if, he added, "the line here marked out were the best possible, which I fear it is not." In short, his predecessor had merely "the fastidious eye of pretended taste" and had presumably been rumbled by Simpson. So Repton found it expedient to criticise the luckless improver. He did so, however, in Picturesque terms, and though he only used the word once in the Babworth red book ("a picturesque cold bath"), it is tempting to see the influence of the conversation at Ferney in his critique.

Thus, there are mild suggestions for hedgerow removal, but the large park he might have so easily have made by removing the hedgerows is set aside as "an object of future attention" in the red book. Instead, he recommended plantations of trees, not to create a park, but in pursuit of "the idea of amply clothing the forest with wood to the west." The key word here

previous pages: In his red book, Repton condemned the ambitious lake and park that had been proposed for Babworth. But he remained involved until the end of his career and the lake was extended over this ground, and the parkland enlarged, in his day.

is "forest"—that is, land recently enclosed from the ancient forest. Forest is Picturesque; it is something we engage with not rationally but emotionally because it is unpredictable and carries the associations of a lawless wilderness, untouched by the hand of man.

The Brownian lake is also dismissed, and Repton suggested instead a rivulet to issue from the Cold Bath, as happened downstream of the pool at Ferney, "and from thence it may occasionally meet the walk in the course of its progress and with very little management always appear alive." Here and elsewhere Repton preferred a stream to a placid Brownian lake. When "enlarged and dug to its gravelly bottom," it would animate the scene and "the water might rattle over some loose stones" and engage Mrs. Simpson with its sound.

Price and Knight demanded that landscape should be composed like a picture if it was to be Picturesque, and hence that there must be objects for the eye to focus on. At Babworth, Repton presented a series of these: besides the Cold Bath, there would be a root house with a suitable rustic wicket gate, an ornamented village, and a windmill. The completed landscape had two further buildings in it: a Swiss Cottage on the east bank of the stream and Lakehill Cottages on the slope a little below it. Furthermore, in the most sustained piece of argument in the red book, Repton was to describe the view from the hall exactly as though he were painting a picture (as Price and Knight required): the birch and alder on the valley floor were to remain to break the line of the horizon; one or two trees carefully placed were to conceal the red roofs of the village; the windmill and road were to provide cheerfulness and give animation to the whole; and the village, partly hidden by the shrubbery, was to show itself by the smoke of its chimneys "curling in fleecy folds against the opposite wooded bank." Repton was never again to describe a complex landscape in this way; instead, we shall see that by the time he came to Rûg, he had become impatient with

opposite: The Bath House appropriates the lake, and the footbridge domesticates it.
above: The Swiss Cottage has been through many changes; it remains a private place removed from the centre of the estate.

the whole idea that landscape gardening must rely on a knowledge of landscape painting.

Even at Babworth the modest experiences that Repton proposed (the Cold Bath, the village) did not only engage the eye's attention, rather they offered engagement and interaction with the world. The touch of water and the smells, sounds, and conversation of the village, which lined the road immediately south of the hall, could be made the objects of a short excursion. By ranking Babworth "below the cumbrous magnificence of irksome grandeur," Repton was ensuring that it would remain approachable, close to the village rather than isolated, connected rather than remote.

When it came to the architecture of the hall, however, Repton was to recommend unity over any Picturesque irregularity, so he used his greenhouse and the adjacent aviary to reinforce the symmetry of the south front, commenting that, "There is nothing so pleasing to the mind as a unity of design which consists in a perfect agreement of all the parts to the

opposite: Repton contrasted the domestic idyll (Mrs. Simpson singing, Mr. Simpson on the 'cello, and himself on the flute) with the wind-swept misery that the lake proposed by his predecessor at Babworth would bring.
above: The scale of the parkland, proposed before Repton's arrival, is best seen from the roof of the hall.

whole, nor is there any thing at which the genuine eye of taste is so much offended as at the incongruous mixture, or misplaced introduction of embellishments." At Honing in 1792, unity remained an essential and self-evident requirement for the landscape of taste: "The first principle of a good composition is Unity of design." It was an idea that he was consciously to reject later in his career.

Given the radical nature of Repton's proposals for the park, the water, and the village, it is not surprising that he should claim that the ideas in the red book were actually his client's: "I am indebted for most of the ideas to your own good taste and observation of the spot." To judge from the red book, the reader to whom Repton principally addressed his ideas was Simpson's first wife Henrietta. Repton is wrongly accused of an obsequious literary style, but at Babworth, he came sufficiently close to it for us to suspect that he was a little in love with her: "I now see her in my imagination leading the honey-suckle round the cottage door and covering the humble thatch with mantling ivy—my fancy presents her passing with you through this new created scene amidst the cheerful smiles of gratitude and love."

She was to die the following year, aged thirty-two, and Repton designed her memorial in Babworth church—perhaps she was already ill and this may have encouraged him to concentrate on those parts of the landscape that could readily be reached on a short walk and to introduce the greenhouse and aviary that were to unify the parkland front of the hall.

Finally, we find that whatever the impact of his disagreement with Price and Knight, Repton was still trying to formulate rules for landscape that would accommodate both Brown's work and his own. At Babworth, he made an early foray into the theory of approach making, which hardened into eight principles at Lamer in 1792, and which remain persuasive today. They can be reduced to two notions: that an approach should appear to be a simple, direct road to the house, and that the first view of the house should be a "burst." Repton was pleased enough with this account of Babworth to refer to it in five other red books.

above: The beeches (left) and yews (right) planted in accordance with Repton's proposal were allowed to grow freely and take their natural form.

above, left and right: In 1795, Repton added to his red book the proposal that the walled garden should be moved across the road and into the parkland, where the woodland that screened it would also protect the hall from the north wind.

COBHAM HALL

— KENT —

CHAPTER THREE

COBHAM HALL

Repton's first visit July 25, 1790,
red book in December of that year

LITTLE OVER A year after Ferney, Repton was commissioned at Cobham Hall, Kent. Like Babworth, Cobham had a Brownian scale; it had also a magnificent Tudor Hall and a sympathetic client, Lord Darnley, who was ready to spend.

Cobham Hall's parkland is best appreciated from the foot of Williams Hill. Looking north, it falls gently to what Repton called a "dry valley" in which the Hall stands, with three small hills to protect it to the north (Windmill Hill), northeast (Peggy Taylor's Hill), and west (Ashenbank Wood).

When Repton arrived in 1790, great works were already afoot. Capability Brown's executor, Samuel Lapidge, had the commission, and the extensive deer park that ran out south and east of the Hall, had been planted in Brownian style, with clumps and single trees. The Pumphouse, by Brewer's Ponds north of Windmill Hill, was up or under construction, designed by Lapidge to boost the supply of water to the Hall and its gardens, which included a Greenhouse, Lady Darnley's Flower Garden, and "America," devoted to North American herbaceous plants.

In the middle of the ancient woodland on Williams Hill, the Darnley Mausoleum was close to completion, to designs by James Wyatt, following the instructions of the 3rd Earl.

Cobham Hall itself was also in Wyatt's hands. His stable wing, immediately east of the Hall, went up in 1790, and while Lapidge was taking down the old farm buildings and stables, Wyatt would have been engaged with the construction of his model farm (Lodge Farm) south of the "dry valley."

Repton's red book for Cobham reads as a carefully worded attempt to replace Lapidge, while preempting any criticism that Price and Knight might make: "... in the rage for picturesque beauty let us remember, that the Landscape holds an inferior rank to the Historical picture, one represents Nature, the other relates to man in a state of Society." In that single phrase, Repton encapsulated his position with regard to his dispute with these two critics of his. The French Revolution and the Picturesque Controversy were barely out of the starting gate, but Repton had already grouped together "the rage for picturesque beauty", Jean Jacques Rousseau, and the natural man (who followed wherever his sensibilities and instinct might lead), and had conferred on them an inferior rank to that of the rational "man in a state of Society." He asserted that "true Taste in Landscape Gardening as well as in all the other polite arts, is not an accidental effect operating on the outward Senses, but an appeal to the understanding which is able to

previous pages: The magnificent Cobham Hall; Repton and his sons refined the symmetries that it now displays.
opposite: The 3rd Earl, Lord Darnley's father, commissioned the Mausoleum because the family vault in Westminster Abbey was full.

H Repton delin

opposite: While the hall was not painted as Repton recommended (above), by 1801 it had been, in his words, "brought to the same tone of colour by sprinkling the walls to imitate the weather stains" (below). below: The South Front of Cobham Hall faces down towards Wyatt's Dairy.

compare, to separate, and to combine the various sources of pleasure we derive from external objects, and trace them to some pre-existing causes in the Structure of the Human Mind." This argument had found favour at Babworth, and Repton still took it for granted that all educated gentlemen, such as himself, were rational, and all shared the same good taste.

Despite these protestations, Repton continued to nod towards Knight and Price in what he actually proposed. Lapidge had drawn up plans for the East and West Courts at Cobham and for West Park, and over a period of twenty-nine days in 1789, he had discussed with the earl the tree plantations and the pleasure ground on Windmill Hill. He is also likely to have made the Brownian proposals for a lake, for a sunken approach from the south, and for moving the kitchen garden away from the Hall.

If Repton were to displace Lapidge, it would help to distance himself from Brownian influence. Thus, he admired the "dry valley," arguing that by extending the pleasure ground on Windmill Hill to take in Brewer's Ponds and constructing a viewing platform on the Mausoleum (with a view to the Medway River), he could bring water into the landscape with more efficacy than by excavating a lake. His critical comments about sinking roads to hide them, his simpler approach from the north, his recommendation that the kitchen garden be left next to the Hall, and his plan for a formal garden could all have been alternatives to Brownian proposals made by Lapidge.

Repton made reference to his red book for Cobham Hall in at least twelve other red books. He also wrote about Cobham in three of his published works (*Sketches*, *Observations*, and *Fragments*). He worked there throughout his career, and his sons continued at Cobham after his death. It is the only landscape to capture every twist and turn in Repton's career, with numerous cottages, a model farm, the "enchanted ground" of Windmill Hill and formal gardens. His Cobham red book was also the first in which we

overleaf: Capability Brown sought to conceal road surfaces through parkland. The sunk road to the house from the south seems to have been constructed, or reformed, by his foreman Samuel Lapidge, shortly before Repton's arrival.

can see a split in his proposals between those that appealed to reason (Brownian sense) and those that sought to elicit an emotional response (the Picturesque sensibility of Price and Knight).

It was good sense to suppose that landscape gardening could be reduced to a set of rules. So, the comment Repton made at Babworth—that where there is a choice of approaches to a house, "the man of taste and the gentleman will take that which is the most beautiful, and the clown will perhaps whistle along the other"—was polished into a rule for Cobham. "If there be two roads of nearly equal length, and one be much more beautiful than the other, the man of taste will certainly prefer it, while perhaps the Clown insensible to every object around him will indifferently use either."

It was good sense to colour the Hall to make it look uniform, as Repton had done at Ferney and Babworth. A haphazard collection of buildings could destroy the ensemble of house and landscape—and in his red book for Bayham Abbey (April 1800), he cited Capability Brown's authority for this: "A red brick house puts the whole valley in a fever." Equally, the garden walls and the offices (that is to say the service buildings) and the stables should be hidden, so as to keep them subordinate to the main building.

It was also good sense to return at a later date to set out "the Drives within the Park, and Ridings in the adjacent Country"—Brown would wait until all other works had been completed to be sure to find "the most beautiful, the most interesting, or the most curious" features of a place. These "Ridings in the adjacent Country" were also very much in Brown's style. Finally it was sensible to bring the north approach further west so as to run through the park: Repton did not want it to be too long, but he preferred the more natural Brownian serpentine to a straight approach to the Hall. However, by moving the approach out to the west, he also created what he would come to call a burst (in which the first sight of the house is withheld until it will make the greatest impression). His approaches were to become a means of conveying a series of emotional impacts. This was a move towards Picturesque sensibility. He was to describe the effect in his red book for Nacton (1792): "Every pleasure derives additional force from being unexpected; so I would in this instance give no previous warning of what should burst upon the sight. I would conduct the walk thro' the plantation to a spot ... from whence in a single moment should open one of the finest scenes in Nature."

Scenery at Cobham had become more than a greenish backdrop; it was now a particular piece of countryside. Repton wished "to enrich the scenery"

opposite: Repton inserted George Samuel's watercolour of "The Four Sisters" into his red book.
above left: In 1804, Repton transformed the fishponds and Lapidge's pump house.
above right: Repton's sweet chestnuts crown Peggy Taylor's Hill.

above: The Reptons' South Lodge is in the 'natural' style of William Wilkins senior.
opposite above: James Wyatt's Dairy (1797) had a thatched milking parlour next to it, designed by Repton's son George Stanley, in 1798.
opposite below: Repton converted the greenhouse into an aviary in 1797, adding its neoclassical façade.

on Williams Hill "by such trees of the Funereal kind," as would create a landscape painting with the Mausoleum as its focal object; hence, also his notice of the "very beautiful scene" that St Mary Magdalene Church made in West Park.

Avenues, on the other hand, did not make good scenery. Repton's antagonism was to surface again at Honing, but he attacked them at Cobham for a fault to which he had drawn attention at Langley Park in 1790: "One great mischief of an Avenue is, that it divides a park and cuts into separate parts, destroying that Unity of lawn or wood which is so necessary to please in every Composition." Here he condemned "the dull repetition of equi-distant trees" as architectural and artificial.

Repton often referred to this discussion of avenues in other red books while he was trying to establish what he wanted "landscape" to express. The word had strong associations with painting; a "landscape gardener" was someone who made gardens look like paintings. In 1791 at Thoresby he had used the phrase of Brown, but by calling their profession "landscape gardening" he made himself vulnerable to the attacks of Price and Knight. By 1798, in his red book for Attingham, he was desperate for a redefinition: "I have been accused of great presumption in having arrogated to myself, the title of Landscape-gardener, but Mr Price confines his sense of the world landscape to pictures instead of that original and more extended signification of the word defined by Dr. Johnson 'a region or prospect of a country' from which the picture is made."

Repton was also ambivalent about old trees in his early career, but here he made a wrong move: after praising the "VENERABLE OAKS" (his capitals) at the Duke of Portland's Welbeck Abbey, he offended Lord Darnley (an acquaintance of the duke) by showing too little respect for his veteran oaks at Cobham. Darnley may well have been influenced by Price and Knight who would praise the Picturesque character of "limbs of huge trees, shattered by lightning or tempestuous winds,"[1] and would criticize Brown, for the

above: Repton's pleasure ground on Windmill Hill is bounded by a conduit that brought water to the house.

instruction "that every thing which indicated decay should be removed."[2]

These attacks caused Repton to reconsider Brown's rational treatment of old trees. He therefore revised the red book to spare "the great avenue towards the Church ... as a specimen of the sublime in ancient gardening," and commissioned watercolours from the painter George Samuel (*c.* 1760–*c.* 1823) to show the "astonishing Chestnut tree called the four sisters," and other old trees at Cobham.

As a further penance, perhaps, to Price and Knight for the terraces he had destroyed at Ferney, Repton retained the straight terraced walk that runs above the north side of the hall. He introduced the idea nervously: "... as an appendage to this ancient mansion I would prefer the broad and stately mall along a strait line of terrace, to the too frequently repeated waving line of Beauty ... I do not recommend the Terrace as an object of beauty, but of convenience." However, after he had repeated the design at Cassiobury, he could report with delight in his red book for Montreal (1812) that the Prince of Wales had told him that "he was so much pleased with the good Taste of the Mall I had made at Cashiobury, that he resolved to consult the person who had in these times dared to advise a straight gravel walk."

The pleasure ground lies above this terrace. It has a gravel drive around the perimeter, wide enough for a pony and trap, while grass walks wind their way through the interior and into a modest quarry. The acreage was too great and too intricate to mow, and Repton proposed to manage it as a sheep walk, using nets or post and rail to keep the stock out of the shrubberies. His design was in the Brownian tradition of a grove, which in turn derived from the idea of an Arcadian sheep walk, with forest trees, particularly oak and sweet chestnut, hanging over grassy glades and a floriferous understorey.

Repton may have had the "Enchanted Ground" of Ferney in mind again—a place made magical by its otherworldliness. Repton took from Brown a regard for landscape gardening as a form of transformative magic.

above: The pond in Repton's pleasure ground was excavated in 1805. The sheep would drink there, completing a picturesque Arcadian scene.

above: The complaint that avenues divide parks had been raised a generation earlier by Horace Walpole. There is a good deal of truth in it. opposite: The parkland and woods at Cobham are studded with ancient trunks, long dead but left to provide a habitat for wildlife.

HONING HALL

— *NORFOLK* —

CHAPTER FOUR

HONING HALL

Repton's first visit March 4, 1792,
red book in May of that year

Thomas Cubitt acquired Honing in 1784. He died in 1788, and his newly married son, also Thomas, immediately invited the young neoclassicist architect John Soane (1753–1837) to work on the house. Cubitt seems to have had plans for a new walled kitchen garden on the moated platform, apparently Tudor, which survives in the woodland north of the house, and parkland extended over the relatively flat, understated countryside of his estate, running south for three quarters of a mile to the beautiful church of St. Peter and St. Paul's at Honing village, north to the ridge of Rabbit Hill, and west to Brick Kiln Grove.

In 1792, however, Cubitt commissioned Repton. There is a suggestion of Thomas's ambition in Repton's admission that "The Situation of Honing, surrounded by its own territory, would make it possible to extend the Lawns," but Repton was immediately to crush this ambition in the rider that this extension would run "far beyond what its character would make advisable." Instead, Repton confined himself to modest proposals for a small park to the south, to take in "the hill field belonging to Lord Oxford," with "a keepers lodge on the top of it, which might take the form of an old tower embosomed 'high in tufted trees.'"

In the face of Honing's potential, Repton's proposals read as no less quixotically small-minded than the advice he gave at Babworth; yet, as at Babworth, his ideas largely found acceptance with his client, both in the landscape and at the house where, rather than further aggrandising it, Repton recommended inserting a string course to make it seem less tall, less grand—a solution to which he gave a more thorough explanation the following year, in his red book for Trewarthenick. Here he wrote: "A line from the cornice of the wings, should be carried thro' the front as a facia to counteract the height of the building, which is too great for its length ... I conceive the reason why, a cube is more unpleasant in a building than any other form; may proceed from the eye having, no guide to call attention, either to the height, or the length." His idea was adopted—but even this most modest of gestures would not have endeared him to John Soane, had he got to hear of it. Indeed, Soane subsequently had little time for Repton, preferring to use the less talented John Haverfield, the Royal Gardener.

Repton also advised against moving the walled garden; instead, he proposed shortening the east wall so as to open a view from the parkland to the house. This was done and a pond was dug in the parkland,

previous pages: Repton settled the house into its wooded backdrop, hiding all trace of the walled garden and offices.
opposite: Repton omitted description of the moated garden where the allées of lime trees were planted after his time.
overleaf: The landscape pivots around the tower of the church, which stands directly in front of the hall.

above: Soane may have added a bay to the house, but Repton domesticated it.

right: With modest planting, Repton composed this tranquil pastoral scene, rising gently to Howard's Hill and animated only by his sinuous drive.

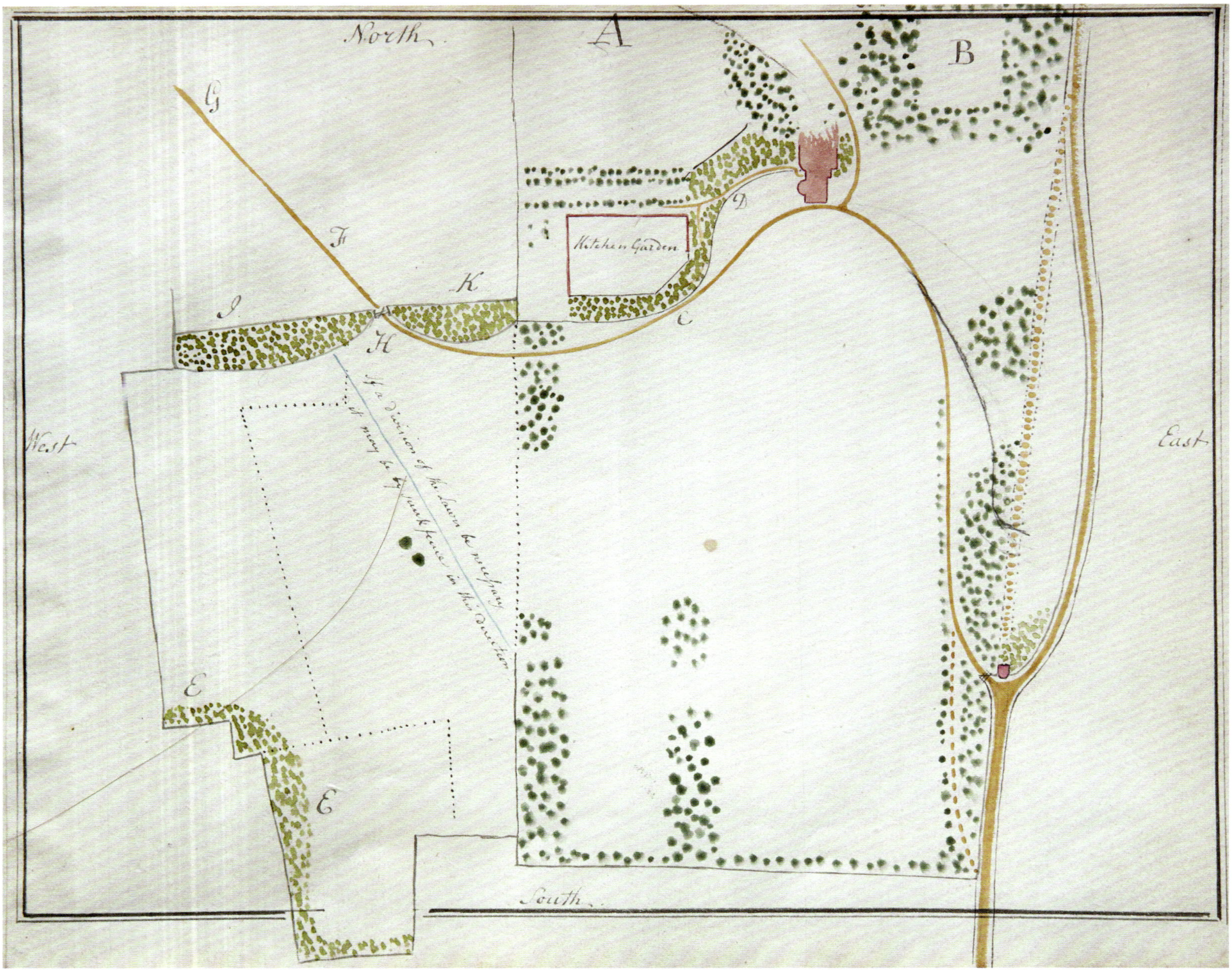
North
A
B
G
F
K
I
H
D
C
Kitchen Garden
West
East
If a Division of the Lawn be necessary it may be by sunk fence in this Direction
E
E
South

though this is too small to have contributed to the landscape, save as a reflective surface in views from the screen of trees that he had planted on its far side and a place where cattle might gather and so animate the scene.

Repton's refusal to extend the parkland at Honing was, however, his most striking proposal. His argument—"however it may be the fashion of Norfolk, to throw down hedges, and unite Lawn with corn lands, I must deem this false taste"—incorporated two principles. Throwing down hedges to extend the parkland and allowing arable land to be seen from the parkland were equally condemned, and Repton managed to imply that Norfolk (and his clients) were both behind the times. He was to curtail his client's ambition for a large park at Sarsden in similar terms, and at Honing, to demonstrate the national reach of his ideas, he referred to three further sites—Hanslope, Wyddial, and Castle Hill—where he had insisted that farm and park should be kept separate. Ironically, Repton did want to create the appearance of a greater parkland by cutting down the avenue of oaks on Long Lane, which runs down the east side, and laying the fields beyond to grass. The Cubitts had just planted the oaks, however, and deemed this a step too far.

It would be easy to conclude that Repton simply hated arable land and cornfields, but this would be a mistake. He could have pushed any arable fields out of sight entirely by making the parkland more extensive, but he chose not to do so. It would be more accurate to say that Repton aimed to keep as much land as possible in arable, his only care being to conceal it with planting, so as to give an appearance of extent to the domain. In 1791, at Glemham in Suffolk, he had already pointed out that in flat country "A large extent of park in reality, is not necessary, but it is essential in appearance."

The timing of this advice also gives the lie to any idea that it was the increase in the corn price after the outbreak of war with France in 1793 that persuaded him to create the appearance of parkland while retaining as much land as possible in arable (there was no prospect of war in 1792). Instead, we should accept that Brownian design did have the creation of an extensive tract of grassland at its heart, and that Repton had been at odds with that idea since the start of his career. By leaving the existing clumps in arable fields, as he had proposed at Brondesbury in 1789, he would "not only ... improve the several landscapes as seen from the home=inclosure but to add consequence, [give] the idea of more extended domain." His alternative—which he threw into his red book more or less as an aside—was to ornament the farm "by neat sweet briar hedges, and broad grass headlands" so as to make "a beautiful detached object." He clarified his position with the comment that "we delight in a ride or walk thro' a well dressed farm, and have pleasure in looking on a distant prospect of a richly cultivated Country, but we are disgusted with the nearer view of the same objects from the windows of a mansion where elegance or affluence are the prevailing Characters."

If a *ferme ornée* can be defined as an ornamented walk around a field, then Repton had a *ferme ornée* in mind. However, he distanced himself from the famous *fermes ornées* of the mid-century, such as William Shenstone's The Leasowes in the West Midlands: "If the yeoman with two or three hundred pounds per annum, presumes to destroy his farm by making what is called a ferme orné, he will absurdly sacrifice his income to his pleasure." He was to elaborate his idea in his red book for Aston, Cheshire, in 1793 with his plan for an "Experimental Farm":

> *It is common in Hampshire and indeed in the neighbourhood of many forests, to divide the inclosures of a farm by rows of copse-wood and timber, from ten to twenty yards wide, and at a little distance these rows appear united, and become one rich mass of foliage. This kind of subdivision I should wish to be everywhere adopted in the experimental farm at Aston, the advantages of this measure will be,*
>
> *1st. Shady and pleasant walks thro the farm.*
>
> 2ly. *It will afford shelter to the corn, and to the cattle that are grazed on the farm.*

opposite above: Repton's plan shows his proposals for remodelling the kitchen garden and screening it.
opposite below: From Repton's "burst" at the south entrance to the park, the hall is still seen as he intended it.

*3rd*ly. *It will give the whole at a distance the appearance of one vast mass of wood.*

*4th*ly. *It will make an admirable cover for Game, and lastly if it should ever hereafter be thought advisable to extend the lawn, it will furnish ample means of leaving many handsome trees to remain single or in groups on the lawn.*

At Honing, too, Repton was to include "a copse or furze cover for protecting game"—shooting and hunting might be amongst the greatest pleasures of country life, but could not take place in parkland, as would become apparent in his red book for Moggerhanger. He had made clear this difficulty in 1790 in his red book for Haling, claiming that no "Gentleman in Sporting" would ever pursue game over the paling of a park, and that "to rear or preserve a herd of game" required "an extent of territory" greater than might be found "within twelve miles of the Capital." This emphasis on the land outside the park was sustained in Repton's proposal that the approach should run "thro' an ornamental part of the farm." He had proposed the idea of "Ridings in the adjacent Country" at Cobham, and the Cubitts adopted it here in modified form with the Loke, Repton's "coach road," that runs through the estate but remains out of sight from the house and pleasure ground.

So, what was Repton trying to do at Honing? We have seen at Babworth that his emphasis on comfort

previous pages: Rabbit Hill is only marginally higher than the hall, but its clump gives it more of a presence.
above: Repton hid the kitchen garden with planting.
opposite above: On Howard's Hill, the Reed House stands on the site proposed by Repton for his prospect tower.

and convenience had nothing to do with the social position of his clients, who included not only titled and landed families, but also many of the most influential men in the government and the City of London. There was room for him to make a great park at Honing. He chose not to. Nonetheless, he felt the need to make the argument again: "... convenience ought to take the lead in a constant residence, since picturesque effect is too dearly bought at the price of comfort."

Repton was to refine the argument at Courteenhall, when he proposed for Sir William Wake "an elegant, convenient, and very respectable family seat; possessing a degree of magnificence compatible with comfort, but not aiming at that which belongs rather to useless Ostentation." Eventually, neatly enough, he was to associate inconvenience and discomfort with old houses like Rûg: "I am however restrained from indulging to its full extent my veneration for Antiquity, by reflecting, that modern comfort and convenience are the first objects to be consulted in the improvement of a Modern residence."

Had he been asked directly why he gave comfort and convenience precedence over beauty, he might have replied, at this date, that to appreciate beauty you first have to be comfortable. Later in his career—and particularly at Endsleigh—he might have said that comfort was a necessary component of beauty. Although he may not have realised it at the time, this was an argument that he came close to making at Honing: "... a stranger may easily decide, whether a handsome waistcoat is best shewn with the coat buttoned, or unbuttoned, but the wearer only can tell which is most comfortable to himself, thus it is not always possible to make what is most beautiful, also most convenient, but it is seldom difficult to render that which is most convenient, also in some degree beautiful."

If the criticism of Brown, which he must have heard from Price and Knight at Ferney Hall, gave Repton a first reason to turn away from the Brownian model, the need for convenience and comfort at Babworth and Honing might have given him a second. At Moggerhanger, we shall find a third.

overleaf, left: Repton wrote of the avenue on Long Lane: "There is sometimes a great degree of importance given to a place, by planting the high road in the form of an avenue leading to the park."
overleaf, right: The Loke is the "coach road" discussed by Repton that serves the estate west of the hall.

1792–1794

EARLY MATURITY

Tom Paine regarded William Pitt the Younger as representing the passions, while he and other revolutionaries represented reason. This notion would have confused Repton. He thought he was representing reason against the passions of the revolutionaries. But Edmund Burke would have agreed with Paine. He lamented the passing in France of "the age of chivalry"; with the accession of pure reason, represented by "sophisters, oeconomists; and calculators ... the glory of Europe is extinguished forever." He claimed that civilisation and the love of country began with an irrational love of property and, while the survival of society might depend on rational behaviour, the "complicated mass of human passions and concerns" and the "refractions and reflections" of primitive man, must instruct that reason.[1]

Then, Louis XVI bungled his flight on June 21, 1791, and inflamed his countrymen. The hard-line Franz II (1768–1835) took the throne in Austria and immediately sought an alliance against France with Prussia, Russia, Saxony, Holland, and Britain, in order to restore Louis to the throne.

France's National Assembly responded, declaring war on Austria in 1792. When the repressive forces of Austria and Prussia invaded, France was kingless but had a population four times the size of Britain's. With a fierce revolutionary zeal and a determination to defend the land over which they had just assumed control, the French defeated their enemies at Valmy on September 20. The Paris mob went on to celebrate with a massacre of Royalist sympathisers, and the National Assembly signed what was to be the death warrant of the Revolution, though not until another twenty years of mayhem had passed. Theirs was the Decree of November 19, 1792, declaring France a republic and "offering help to all those nations who desired to overthrow their kings." The crowned heads of Europe could not allow such an offer to go unchallenged.

William Pitt's "you've never had it so good" speech of February 17, 1792, could not have been less timely. Eleven months later—too late and out of touch—he called on France to renounce "all ideas of aggrandisement" and "to confine herself within her own territories." March 21, 1793, saw France guillotine Louis XVI. Pitt's response of "No event can be looked for more desirable than a destruction of that system, which at present exists to the misery of France, and the terror of Europe,"[2] brought Britain into a war that, with one brief pause, was to last for the rest of Repton's career.

All was not well at home either. Riots in 1791 in Bristol and Shropshire were fuelled by rising prices as much as by the Revolution. Birmingham, memorably, took fire to the dissenting scientist Joseph Priestly, and the two countries marched side by side into tyranny: "We declaim against the French, & we imitate them in their most horrible atrocities."[3]

During the spring of 1793, Paris established its Committee of Public Safety. On the second of June, Maximilien Robespierre took power and spawned the Reign of Terror. Pitt responded with the Alien Office, ostensibly to root out foreigners, in practice to control a network of spies and informers across the country: "... the most powerful means of Observation & Information ... that was ever placed in the hands of a Free Government."[4]

Pitt had turned despot. In two and a half years, he had brought down the lid on political freedom and clubbable eighteenth-century society. The revolutionaries were atheists. Therefore, the church became a bulwark of the national interest. Its Pittite congregation prepared for self-sacrifice, social cooperation, and war. Suddenly, a retired life, removed from the levers of political oppression, might attract the liberal conscience.

In this unlikely period of turmoil and division, at Port Eliot, Ston Easton, Moggerhanger, and Mulgrave Castle, Repton found his own voice as a landscape gardener.

opposite: Repton played to the Picturesque character of the land at Mulgrave Castle.

MOGGERHANGER PARK

— *BEDFORDSHIRE* —

— *CHAPTER FIVE* —

MOGGERHANGER PARK

Repton's first visit May 28, 1792, first red book in August of that year, the second, "The Beeston Leasowes," in 1798

By 1792, Repton's client, Godfrey Thornton (1737–1805), was playing a considerable part in national life. He was a partner of his own bank and in 1772, at age thirty-five, had been appointed a director of the Bank of England. In due course, he would become governor. His family had owned Almond Farm, as it was then known, and 1,024 acres of Bedfordshire clay for more than sixty years. Their small house stood—and still stands—at the northern end of a slight rise, the ground falling gently away from it to the northwest and southeast.

Thornton was a close friend of William Wilberforce, an evangelical Anglican and member of the "Clapham Sect," a group devoted, among other things, to the abolition of the slave trade. In short, Thornton was occupied in London and intended Bedfordshire for occasional visits. It was an unlikely site for the first complete work of Repton's maturity.

Thornton took over Moggerhanger from his brother Robert in 1784, and by 1789, he was building, among other things, garden seats, a greenhouse, and a walled garden, while the house was in the hands of John Soane (surveyor to the Bank of England from 1788). Their collaboration produced chaos: Soane's new drawing room was next to the front door, which made it "almost impossible to preserve an air of neatness" for "the hall door requires an expanse of gravel for coaches to turn upon, and is exposed to the occasional defilement of Horses waiting at the door, and immediately under the windows." Still worse, the traffic between the farm on the north side of the house and the walled garden on the south side necessitated continual noise, smell, and interruption in front of the windows. Furthermore, the house overlooked fields to the south that were not owned by Repton's client, while the 80 acres to the north that Godfrey had earmarked for a park were largely out of sight. Repton's friend John Byng summed up the challenge in 1794: "Having a wish to see what alterations and improvements Mr T was making at Muggerhanger, we rode three miles further to that place which is truly a dismal hopeless spot ... it wants water, soil, and timber."[1]

Repton immediately claimed control of the work:[2] "... altho' Landscape Gardening is often considered as more subject to fashion or caprice than any other Art, it is not less reducible to rule than Architecture, or Painting: since True Taste in all the Arts must be founded on reason and reflection." We know that Repton visited in 1792 and 1798, and given the complexity of the design, it is likely that he visited many more times.

previous pages: Lowland Bedfordshire has a heavy clay, good for Brussels sprouts and bricks.
opposite: A tree stump survives from the row of trees planted along the top of the ha-ha.

above: Repton eventually had the entrance moved away from the east front to the north, so giving him space to make a small ladies' garden here.
opposite: Repton would use the columns of a portico or, here, the treillage of a veranda, to break a broad view into separate vignettes.

Repton was troubled by the rules that he had begun to compile at Cobham. He had divided houses into five groups, but Moggerhanger did not fit the model. Later in his career, he might have been happy to describe it as a cottage, but the best he could manage in the early 1790s was "a retreat for sporting." Moggerhanger was a place where Thornton could fish, hunt, and carry out improvements without the weight of responsibility that attended his life in London.

Repton began the project by introducing a degree of symmetry to the Lawn: planting at Park Farm to balance the screen planting around the walled garden and establishing entry points, such as the end of the belt along the Northill road, the Budna road junction, and Soane's lodge on the Bedford approach, at approximately forty-five degrees from the main axis. By so doing, he hid the points of entry from the windows while displaying two fronts of the house at once on the way in, thus adding "great importance" to the house, as he wrote at Lamer in the same year.

He then gave the ground increasing polish as it came closer to the house. The Bedford approach ran straight across fields that he left in arable, past Soane's lodge, then into the pasture and hay meadows of the parkland, followed by the more intricate planting of the Close, immediately north of the house, and finally reaching the house itself and its gardens. The same gradation ran out to the southeast, from the veranda, across the polished landscaping of the Lawn, down to the Northill road, and up again towards the arable land of Beeston Leasowes; there, he provided a "preparatory effect," as he was to describe it at Courteenhall. At the top of the public road from Budna, there was a first glimpse of the house, caught between two of the clumps on the Beeston Leasowes, designed to alert the traveller to the true "burst" three hundred yards further north—a composed view of the house, standing above its pleasure ground and the polished parkland of the Lawn.

Repton had recommended a veranda for the house in his second red book for Moggerhanger (1798). The veranda was not installed until 1811 (not an unusual delay for Repton), and cedars and a copper beech were then planted along the sunk fence off the east front of the house. These trees shielded the dining room from the glare of the afternoon sun (one of Repton's lasting concerns), and were carefully planted so that each one stands behind one of the uprights of the veranda when viewed from the threshold of the house. The combination of a tree and an upright broke up the panoramic view into a set of separate landscapes. Furthermore, the species chosen had dark leaves and so gave a *repoussoir* effect such as he had already tried at Tendring in 1791: "The situation of the trees immediately beneath the house makes it in a great degree partake of the Landscape, by breaking the view into distinct parts ... "the whole is set in a dark foreground."

This new pitch of precision is illustrated again in his use of short ha-has. He had introduced them at Stansted in 1791, and wrote about them in his red book for Milton shortly afterwards: "The most happy effects may often be produced by a sunk fence opening for a very few yards only." These constructions could be framed by trees to conceal surrounding arable land, a simple, but essentially Picturesque, way to create picturelike scenes. They suited Repton's notion that landscape might be made up of a string of distinct and discrete sensational episodes—an appeal to sensation to which he still claimed to be opposed.

There are three of these short ha-has around the Lawn. The first, along the Northill road, allowed views up to the eastern Beeston Leasowes. The second ran between the house and Park Farm. By felling "a few trees" from the avenue there that "might perhaps be taken away to advantage," he could connect the Lawn on one side of the house with the Close on the other. Finally, in his supplement of 1798, he recommended "... sinking the park pale for a certain distance near the road" to make "the Beeston Leasowes ... appear part of the Park." Framed by two clumps of lime, this view took in Sheerhatch Wood and the "Wood House," a woodman's cottage, no longer standing, that he proposed for the entrance to the wood "... so ornamented by flowering shrubs, & by painting the

opposite: The house is only seen in the last two hundred yards of the Bedford approach.
overleaf: Soane returned for decades to work on Moggerhanger House, steadily transforming the old farmhouse into a masterpiece of neoclassicism.

door & windows green, that it may not appear the humble habitation of a poor Labourer, but rather the reward of some favourite Servant, made comfortable by the elegant attention of the Ladies, who may be supposed to take pleasure in decorating this little Spot. The Cottage is so formed, & so placed, as to catch the light of the afternoon Sun, & become a cheerful feature in the Landscape from the House." He had built the Woodman's Cottage at Blaise Castle in a very similar style by this date.

Each of these sunk fences placed constraints on the form of the planting on the Lawn, and in addition Repton had to accommodate the Thorntons's interests. Like the Simpsons at Babworth, the Thorntons valued agricultural improvement. However, as Repton had explained in his red book for Honing, he had "frequently had occasion to explain the doctrine, of not permitting corn fields to appear in contact with park scenery." Here, therefore, he brought the park boundary in from the Bedford road and planted a screen of trees along it, thus allowing a large acreage to be turned to arable. At the same time, within the parkland to the northwest, he retained three paddocks with their hedgerows, which he liked and described as "forestlike bordering," while complaining that "in other parts ... hedges have been too hastily destroyed." These hedgerows were out of sight of the house, and they were excellent for game. Repton had discussed such paddocks at Babworth in 1790. Bottom Wood too, farther from the house, was to be coppiced to make "excellent nurserys for game, and answer the expectation of the Sportsman

above: At the "burst" from the Budna road, Moggerhanger dominates the intervening land.
opposite above: Repton was to remove the hedge and make this gateway the start of his approach.
opposite below: From Garden Wood, Repton was to make a "peep" view of a neighbouring house, the Hasells.

No. V.

No. VII.

better than lofty woods," again he had made the same recommendation at Honing.

Garden Wood, on the other hand, gave Repton the chance to create another "Enchanted Ground," such as he had found at Ferney. There were to be walks, "bold and easy, often varied, and sometimes intricate ... without being tortured by perpetual wreathing, or rendered insipidly dull by uniform undulating lines (of beauty as they are often miscalled) ... some amply shewing the stems of neighbouring trees, thro' whose foliage the sun beams may chequer the neat mown turf of the winding glade, and others as mere paths, darkly brushing thro' the thick covert of a closely tufted copse ... from winds a shelter and from heat a shade." These walks would have bursts to "those points of view, where some striking feature rewards the expectation." A rustic summerhouse was built near the southeast corner of the wood, with fine views to the north; to the east, towards the Hasells, the great house of the Pym family, there was an Alcove or covered seat; and there were bursts to a so-called "Tudor Cottage" (an aviary, or hen-house), which presented its more Picturesque west side to the wood.

Gradation was not limited to the parkland, but could be used to extend the range of the domain by means of ridings that would break through the boundaries of the parkland and run on into the countryside. As at Cobham and Honing, Repton could propose "an interesting drive ... towards that part of the estate which borders on the River." Thus, he introduced water into the landscape, and "a cottage might be placed with an occasional room, for those who come to fish."

Thornton was also building cottages for his workers and dependents and, unsurprisingly, Repton wanted to incorporate the village as well—all it needed was a uniform garden fence and some evergreens "to hide the more shabby or dirty parts ... [because] the houses in Mogenhanger parish are already in very neat repair, they only want a few honeysuckles or vines to decorate their walls, and some evergreen trees to unite and combine the several detached buildings ... [and] nothing contributes more to the importance of a place, than the appearance of neatness and attention to the comforts of the neighbouring poor." However, the chosen route for the Bedford approach bypassed it, running over the higher ground farther north, to Soane's lodge. This may have been Soane's rather than Repton's choice, but it displays further details of his design. There is a significant sequence from dark to light in the planting of the Close: from oaks (mid-green leaves) to limes (light-green leaves, dark trunks), and then beeches (light-green leaves and light trunks). Another Picturesque effect followed: the approach passed "betwixt the two large trees." Repton regarded these two old elms as "the finest features of the whole place." They were not only Picturesque in themselves, but by running the approach between them, he could also make them frame a burst to the house.

Knight and Price despised Brown's lawns and lakes; they loved old trees and they also believed that landscape should emulate painting and hence required objects to provide a focal point. Repton had already adopted this idea at Babworth, but at Moggerhanger he gave the role of an object to the view his full attention: "Church steeples and windmills ... may be sometimes singly introduced in those little peeps thro' a thicket in the course of a walk," but at any distance over a mile, "five, or five score of Churches, which look no better than barns, can be no object of rational enquiry." Repton's opinion, which allowed for objects to make sense of a "peep" (a narrow picturelike view), but dismissed such objects in a landscape, was exactly at one with an idea expressed later by Uvedale Price himself.[3]

Moggerhanger suggests further reasons for Repton's success with these understated designs. Clients like the Thorntons were the new elite, but they came from different backgrounds and had different values from the old aristocracy. They had less land and their country houses were smaller, so Repton gave them smaller parks. They were Evangelical, they believed in work, they wanted their ventures to make money, and since the French revolutionaries were rationalists and atheists, their Christianity made them patriots.

opposite: The Aviary Yard had been built before Repton arrived, its homely character appropriate for a gentrified farmhouse.

above: Repton painted the disordered view from the east front of the house as he first saw it.

below, left and right: Repton filled in the ponds at the east front of the house and ran his approach between the two large elm trees.

The gently rolling ground of Bedfordshire presents a considerable challenge to the landscape gardener. It would not be met by counting the number of steeples in the view.

COURTEENHALL
— NORTHAMPTONSHIRE —

CHAPTER SIX

COURTEENHALL

Repton's first visit March 23, 1791, red book in March 1793

Sir William Wake of Courteenhall (1768–1846) has already been mentioned. He had been a wayward boy. Samuel Phipps of Ferney Hall then acted as his guardian and left him a legacy of £5,000 in 1792. Repton had first visited a year earlier, and he came with a glowing reputation from Ferney (Repton picked up many commissions by word of mouth, itself evidence of the high value put upon his services). The Wakes for their part were a family that had much to engage his imagination: they were well-established, with a substantial estate and claims to descend from the legendary Hereward the Wake, the eleventh-century Saxon warrior who had led resistance to William the Conqueror in the fen country.

Repton's star was still rising and likely his confidence was at a high, but he may also have been relieved to find himself in Northamptonshire, which was, in his words, "certainly not picturesque," and hence free, perhaps, from the oversight of Price and Knight. The architect Samuel Saxon had already designed the new house and a pair of Palladian lodges. Repton was to advise on their location and on the necessary outbuildings, and then to propose a landscaped setting for the ensemble. He seems to have begun sketching for a red book on his first visit in the spring of 1791. His first sketch showed the old house, but he returned several times, "marking out much of the detail on the spot," and by the time he came to submit the red book in 1793, the old house had gone and "disregard" was scribbled onto the back of the flap that showed it.

He was not convinced by the look of Saxon's design ("The mansion at Courteen hall being intended as a Family residence, and not built merely to please the eye"), and felt that a grander house would be needed to outshine the stables ("to display this whole edifice, and make it [as a stable doubtless ought to be] subordinate to the mansion, would have required a palace"). Nonetheless, the site he chose was near them and near the old grammar school (set up in 1672, when the open fields of the parish were enclosed). The new house was to be built on a level shelf in the middle of parkland that sloped gently down towards the county town of Northampton, five miles to the north (as it happens, Jane Austen's Mansfield Park was a similar distance from the same town).

It seems clear that Sir William had planned to build Saxon's pair of lodges at the turnpike road on the edge of the neighbouring village of Blisworth, but here Repton intervened: his preferred approach was to run in from the northwest rather than from Blisworth, and he may not have wanted Saxon's lodges at all. In his red book for Bayham Abbey (1800), he

previous pages: Much of England's parkland in Repton's day would have been a wildflower meadow, managed for hay.

opposite: Saxon's design for the new house had already been approved when Repton arrived at Courteenhall.

Repton sited the new house on top of the slope to give it a weight to match that of the stables. Later in his career he might have preferred to keep the manor house.

was to condemn such designs as "the miserable little double tea=cadies of modern taste." Whatever his attitude to the lodges at Courteenhall, he may also have wanted to make it clear that their location and design were his business, but rather than upset both Saxon and Sir Charles by confronting their ideas, he involved himself in a lengthy argument.

First, he agreed that symmetrical lodges on the turnpike would have a "preparatory effect" on visitors to the house itself and would announce to passersby that this was "a Gentleman's place." Then, Repton applied the theory of gradation that we have seen at Moggerhanger to criticise this location and proposed instead a thatched cottage, as he had at Babworth—this one with a "rude portico" to nod at the neoclassicism of Saxon's house. Saxon's lodges might be retained but would better answer to his idea of gradation if they were brought closer to the house and built on the west edge of a plantation of trees, designed to shelter the pleasure ground and ensure that the house was seen before the stables.

However, Repton always accepted the straight "direct" approach from Blisworth. This would have been used by tradesmen and the family when off duty. It ran outside the parkland, which Repton wished to contain because "there would be no real advantage gained by extending the park to the hedge near the turnpike," and it would be at one with straight sections of the approaches at Ferney, Moggerhanger, Rûg, Sarsden, and Stanage.

Repton's preferred principal approach, on the other hand, was Brownian. It gave a shorter route for

previous pages: The stable block is a magnificent building and Repton planted to prevent it dominating the house.
above: The walnut and lime tree frame the "burst" to the house from Repton's Northampton approach.

Northampton traffic and showed off the parkland, while preserving its natural character, because it was grassed. He had rejected something of that kind at Babworth, but there it had been proposed by an unnamed Brownian rival. In 1792, at Tatton, Repton had also ridiculed Brown's "aversion of shewing a road" in this manner as "so strange," but at Honing, he had recommended these grass walks for summer use.

Courteenhall offered a fine open view of Northampton from a point some 300 yards north of the old house, and Repton had succeeded with a similar view towards Ludlow at Ferney. The challenge here however, as at Ferney, was to landscape a downhill slope across land not owned by the estate. Repton declared this downhill slope to be "one of those subjects which painting does not well represent," and hence out of reach of Picturesque theory, and instead chose a site for the new house that was closer to the old one.

Repton's view has been blocked today by plantations of trees designed to hide the M1 motorway, but there is an ancient walnut tree directly in front of the front door, with a lime tree behind that. These trees divided the view into two, with a narrow glimpse on the left towards Repton's cottage lodge and a broader sweep of the downhill slope to the right. Repton would seem to have positioned the house to take advantage of the walnut (which also framed a burst from his summer approach), and this makes Courteenhall an early example of a house orientated towards the view rather than square to the slope. Repton was to allude to the idea when working again with Samuel Saxon at Buckminster in 1793: "... contrary

above: Repton sought to resolve the "want of variety and intricacy" in the tree belt that had been planted before he arrived.

Repton retained the straight approach from Blisworth (B), and effectively cut the areas C–F out of the park. Later, he revised the plan, adding an area that he set aside for a flower garden in pencil, and proposed water bodies in blue.

5 N.1.
Kitchen Gardn
School
C
B
D
H
The Turnpike Road from Northampton

to the absurd practice of some Architects, who first build the house on paper, and then make the ground accord with the house; Mr Saxon feels the necessity of adapting the house to the situation." Saxon had a partnership with the architect John Nash (1752–1835) at the time, and the idea was one that Nash was to claim for his own at Castle House, Aberystwyth, which he designed for Uvedale Price in 1795.

Once Repton had dealt with Saxon's house and the lodges, the view south presented a second problem. This was a rising slope with the church of St Peter and St Paul as a focal point, just east of being directly in front of the house (a position that Repton later described as "point blank"). To balance the view on the west side of this point blank, there was what looks like a fold yard with a pond to attract the stock to it, immediately in front of the boundary belt of trees. This belt was not his—in fact, a good deal of work had gone on immediately before his arrival, with the removal of the village, the closure of several roads that would have interrupted the new parkland, and the decision to remove the old house. As for the belt, Repton complained that it was "one of those belts which give an air of confinement by the want of variety and intricacy in its outline." To Repton, though, we should attribute the group of trees planted in front of the belt to break its line, something he was also to do at Brightling. First, however, Repton had to prevent the church from dominating the landscape. He achieved this by bringing his "skreen" (the spelling is Repton's) in front of it. At Stoke Edith in 1792, he had complained that "I have too frequently occasion to regret; not so much on account of the church itself, which may sometimes be rendered a pleasing object in the scene, but on account of those appendages which frequently accompany the church." But whereas the Repton's skreen is dominated by horse chestnut, the church was surrounded by beech. At Courteenhall, he was applying the principles of aerial perspective that we have seen at Moggerhanger. It remained a lively idea for him in 1792, and he mentioned it in a number of his red books for that year. At Tatton, he wrote: "The second kind of perspective is called Aerial ...

left: Though dominant in the winter, the tower of the village church and the rectory beside it are largely hidden by trees in summer.

objects ... nearest are strongly represented while other parts of the scene as they recede become less distinct, till at last the outline of a distant hill seems melting into the air itself."

In short, Repton would have expected the lighter-coloured beech leaves to distance the church. Since beeches also suppress the understorey, they would also have allowed glimpses into the park from the churchyard, an effect Repton had described at Honing ("betwixt whose glossy stems may hereafter be let in distant views of those particular objects whose exclusion may now for a while be regretted") and had used in Dog Kennel Plantation at Babworth.

The red book has obviously been revised a number of times. For example, Repton gave several reasons for not making a lake, yet a proposal for water was included in the text, and ponds were crudely added to his plan, presumably by himself. The supposition must be that the page headed "Of the Water &c" was added at a later date. Furthermore, the pages of the red book are out of order and, in a pencilled note, Repton explained that he had withdrawn his Sketch No. III "til the offices are built or at least the plan decided."

Another later addition was Repton's plan for a flower garden. Its enclosure is shown on a plan of 1814 and survives in outline today. Once again, we may trace the rapid evolution of Repton's ideas—from his questioning the role of formal gardens in his red book for Welbeck Abbey (1790) to his proposal for planting between the wings of Cobham Hall in the same

above left: Tree trunks were to be set up as Doric columns at Repton's lodge.
above right: Repton often used copper beech (grafted in this case) close to the mansion.

year, which he later described as "a flower-garden, enriched with marble statues and a fountain, forming an appropriate frame, or foreground, to the landscape of the park."[1] Courteenhall's, by way of compromise, was neatly placed off to one side, ornamenting the walk between the house and stables and screening the offices, without interrupting the open parkland setting of the house.

Repton's work at Courteenhall reveals a new confidence, illustrated not only by the orientation of the house and the flower garden, but also in his decision to build the house so that it faced directly towards the walnut tree. In some respects, however, he remained locked into the landscape tradition and the past. He retained old trees wherever he could. He showed the new village of Courteenhall on his plan, but did not mention it in the red book as we have it. The removal of the old village had begun some time after enclosure in the late seventeenth century, but was completed by Sir William, and now its replacement was tucked out of sight and was not to form any part of an approach. It consisted of a couple of architecturally understated terraces, each consisting of four houses. Like the estate villages designed by Brown at Milton Abbas and Nuneham Courtenay, these houses have no front gardens but stand against the street.

Whether or not Repton had a hand in these terraces at Courteenhall, the idea of regimented housing for estate workers was one that he had abandoned by 1796 at the latest.

overleaf: The shadows of the trees ripple across slight ridges, a relic of mediaeval farming. Lime trees on either side dominate the house and prevent it from sitting up too high in the landscape.

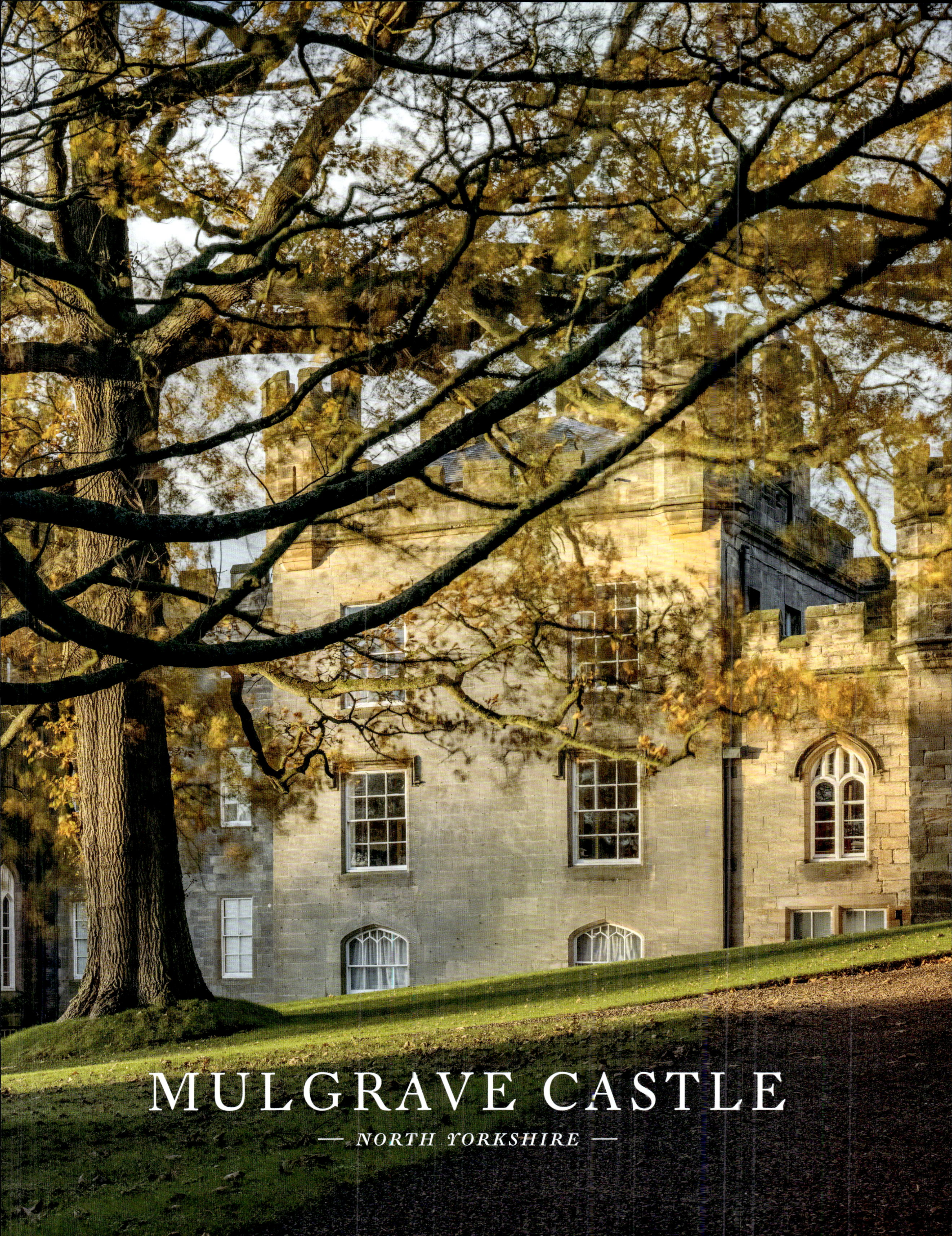

MULGRAVE CASTLE

— NORTH YORKSHIRE —

CHAPTER SEVEN

MULGRAVE CASTLE

Repton's first visit September 3, 1792, red book in August 1793

IF MOGGERHANGER SHOWS how adroit Repton could be at integrating a modest house into the surrounding countryside, and Courteenhall (grander, but the Wakes were only baronets) shows him capable of developing a cut-down version of Brownian parkland, then Mulgrave Castle, on which he was working at much the same time, gave him an early opportunity to show off his skill with a topography that was both extensive and, potentially, Picturesque. The estate is perched on the east coast of England, at the point where the wild moors of North Yorkshire crash into the North Sea. The great house stands above and apart from the sea to the east, and a cross-hatching of steep valleys on its south side. There is no view to speak of to the north where, in Repton's day, the village of Lythe ran up to the back door, while to the west the land rises to the moors.

It could be a bleak, cold place, much as described by Mrs. Lepel Phipps, the mother of Repton's clients, John Constantine and Henry, in 1743: "... a much prettier Prospect than a convenient dwelling, since the Hills are all so steep & the ground so stony and uneven," made worse by the fact that "this country is very open, there is little or no wood any-where." Mulgrave's only small compensation being that "we have a full view of the sea, which from the window seems to wash the very Borders of The Garden, tho it is at a quarter of a miles distance from it."

The sea was to be a compensation for Repton's client. John Constantine Lord Mulgrave was a Navy man, descended from a family with a long and very admirable record of public and military service. He had succeeded to the title on the death of his father in 1775, but continued to have a very distinguished career as a ship's captain, not leaving active service until the end of the American Revolutionary War in 1783, when he became Paymaster of the Forces, Commissioner for India, and a Lord of Trade.

Small wonder that in 1785, finding himself possessed of a substantial income, Lord Mulgrave could commission Thomas White, once a foreman of the late Capability Brown, to draw up a plan for the park. The first of the walks along the steep valleys of the Sandsend and East Row Becks had already been built. A purely Picturesque scene of the water tumbling through the rocks was sketched by the landscape designer Fredrik Magnus Piper in 1773 on his way back to Sweden (Repton painted the identical view in 1792).

White's plan, however, was Brownian in form, with a narrow belt of trees surrounding about 400 acres of parkland on the high ground north of these becks.

previous pages: Mulgrave Castle was Repton's most northerly commission in England.
opposite: Repton's tunnel was built and is still in use.

Lord Mulgrave shelved this plan, brought in John Soane to reconfigure the house and walled garden, and set in motion a very extensive programme of over 60,000 trees to plant the valleys, overseen by Adam Mickle, another foreman of Brown's. The ground would not have been much use for anything else, but the new woodland was sheltered, and enhanced the Picturesque character that Piper had sketched eight years earlier. Even the names of these plantations point to this new rugged aesthetic: Birk Head, Trucky Rock Hole, The Haggs.

It is not clear why Lord Mulgrave should have settled on Repton as his professor in "the Art of Landscape Gardening." He and his brother were adherents of Pitt the Younger and may have heard of Repton from colleagues at Westminster. However, the Mulgrave red book is immediately striking—and unique—for in it, Repton committed himself to make "Mulgrave as interesting an ornament to the north of England, as Mount Edgecomb is to the south." Mount Edgcumbe was the most famous landscape in the west country, in part because of its position overlooking Plymouth Sound, but also for its pleasure grounds (terraces around the house and a series of hidden gardens in various styles) and for the Amphitheatre (a U-shaped valley in which the walled garden sat), the Zig-Zags (a series of drives terraced into the steep slope), and a considerable number of small ornamental buildings, mostly scattered on or around the fringes of a large grass hill behind the house—incoherent, but even today, exhilarating in parts.

Repton claimed: "I have frequently been asked which was the finest Place that I had ever seen, and till I had visited Mount Edgecomb I never hesitated in pronouncing, that the scenery at Mulgrave was altogether the most magnificent." This was not an obvious route to getting the Mulgrave commission, and since he was writing only eight months after his first visit to Mulgrave, these "frequent questions" could only have been put in that time. Who more likely to interrogate him than his client? Lord Mulgrave would have known Mount Edgcumbe very well since

left: A lawn extends east from the house to the Quarterdeck with views across the parkland, known as Barfs, and over the North Sea.

Repton's "after" view from the library shows the village of Dursley with his proposed "conspicuous building." This was never built, nor was the land around it emparked.

N° II

N° II

it overlooked "the Harbour of Plymouth, and the busy scene of shipping and dock yards"—the centre of his professional life since he first joined the seventy-gun HMS Monmouth in 1759 as a cadet under his uncle, the Hon. A. J. Hervey.

It would be reasonable to suppose that Lord Mulgrave had already declared his intention to make Mulgrave a second Mount Edgcumbe and had tasked first White and then Mickle with the job, but found them wanting, and so called in Repton. Repton's difficulty ("the event which caused this opinion to be so long delayed") was that his client died one month after commissioning Repton, and it was Henry, his brother and successor to Mulgrave, who asked Repton to submit the red book that he had been preparing. Repton emphasised his desire to return to Mulgrave for a second review of the property first ("it would require a more recent knowledge of the spot"). He may have known that his mapping of the south side of the landscape was wildly inaccurate, showing the village of Dunsley over six hundred yards northeast of its true position, which made unrealistic his conjectured parkland setting for the village.

The ideas in his red book were largely adopted, however, and there are good grounds for supposing that Repton did go back "to direct the detail of its execution." His plan does not cover the parkland proposed by White, although his "after" painting shows it as planted. The text of his red book was more concerned with the two valleys to the south.

above left: Whitby and its abbey were Mulgrave's Plymouth.
above right: The bridge across Sandsend Beck has been renewed but sits on original abutments.

A southern approach would make an attractive route from the Whitby Harbour and the views to the house from the south remain romantic in the extreme. He could wind the approach along the course of a minor beck, sometimes bridging it, cross the Sandsend Beck "in a romantic dell," and then drive a tunnel through Castle Rigg, the high spine that divided the valleys, "with all that magic of effect that always attends the emersion from darkness to light." So developing a highly Picturesque area of around 60 acres south and west of the Old Mulgrave Castle, with the Hermitage (claimed to have been first built in 1150), the Amphitheatre (the name also used in Mount Edgcumbe's landscape), and the becks, which Repton valued here as he had at Babworth and, in all probability, at Ferney Hall's dingle, with "the prattle of the brooks, and the murmur of the cascades ... solemnly accompanied by the deep roar of ... rolling waves." By 1812, these elements had been further enhanced by the tunnel and Dry Arch, connected by numerous drives, running along the side of the hills, terraced into them and reminiscent of the Zig-Zags at Mount Edgcumbe. For Repton, this was surely an "enchanted ground," such as he had imagined at Ferney and recommended for Cobham. His account of the treatment proposed for the walks and becks in this area might be mistaken for a description of Downton Gorge, the Picturesque high point of the estate of his critic Richard Payne Knight: "The brooks assume

overleaf, left: The paintings by Repton and F. M. Piper are nearly identical.
overleaf, right: The becks that run through Mulgrave are a gift to the picturesque imagination.

opposite above: Old Mulgrave Castle crowns the ridge between the two becks.
opposite below: The Dry Arch was a device for framing a view of the sea in the character of a defensive outwork of the Old Castle.
above: Soane's Mulgrave was a neoclassical box, and Repton's advice that "a crenelated appearance" would be more in character was adopted.

different characters according to the channel which they have worn at different places, and it is the duty of the improver to present this variety under the most striking circumstances of contrast; Walks, or rather paths should be formed in the wildest manner, yet leading without difficulty to the most interesting points of view, and where it can be no otherwise effected, the path must occasionally be cut thro' ye rock, or a descent made by easy steps down the precipice." To this he would add "a degree of confusion and intricacy ... to render the line interesting ... by making it a little difficult to trace its actual course," while making the woodland flow by planting the hills and encouraging "a sweeping continuity of wood, connecting the hills and vallies."

He also made sure to distance himself from the Brownian boundary planting that White had proposed: "If a belt or girdle of plantation were to be made round the skirt of the park, it would seem artificial however varied might be its breadth or outline." Yet, despite his evident commitment to the romantic and Picturesque, Repton still retained many of the ideas we have seen expressed in earlier works; he wished to ensure that the ruins of Old Mulgrave Castle would remain subordinate to the house in the interests of the overall unity of the landscape. He

above: The paths that Repton had imagined can still be traced through the woods.
opposite: Repton's "most interesting points of view" stand out as levelled promontories or mounds.

also proposed to bypass the village of Lythe, with its "ruinous cottages" (the village had gone before the end of the nineteenth century, and presumably the properties were ruinous because Lord Mulgrave was taking them in hand with a view to demolishing them). However, the most conspicuous of all his recommendations now stems from his hint that "the new mansion has been called Mulgrave Castle without any symptom of a Castle form, or even a crenelated appearance to warrant the appellation." From 1804, the architect William Atkinson faced John Soane's neoclassical improvements with an essay in Castle Gothic.

What makes Mulgrave a triumph is not any imitation of Mount Edgcumbe, but the way its wildness remained untamed. Repton's new approach from the south was designed for utility and convenience. It was protected "from every appearance of danger," but remained exciting, dramatic, and thrilling. The exhilarating enchanted ground, the open space of the Old Castle interior, and the woods around—the creation of episodes like the delightful glade of Ash Holm deep in the heart of the woodland, with no sound from its grassy banks but the beck rushing through—these were encounters to stimulate a Romantic imagination.

overleaf: The walled garden offers some protection against the wind from the North Sea.

1794–1796

THE PICTURESQUE CONTROVERSY

In October 1795, Napoleon Bonaparte (1769–1821) crushed a royalist uprising in Paris. In 1796, he retook Corsica, marched into Italy, smashed the Austrians, and annexed the Austrian Netherlands.

William Pitt did not have his country behind him. In October 1795, crowds threw refuse at George III and demanded cheap bread and peace. Pitt responded with the Seditious Meetings and Treasonable Practices Acts, which effectively banned public meetings and stifled political debating societies. The United Irish, a coalition of Catholics, liberal Protestants, and dissenters, took their chance to lead a war of independence, and a French expedition set off for Ireland in support. The attack failed, but with the threat of invasion, food shortages, and economic disruption, it is not surprising that by February 1796, Pitt too was making plans for peace.

Repton was also beleaguered. In 1794, Richard Payne Knight published his poem *The Landscape* to coincide with his friend Uvedale Price's *Essay on the Picturesque*. These publications attacked Brown and, by implication, Repton, who responded to Price and Knight in 1795 with "Letter to Uvedale Price Esq. On Landscape Gardening," making this a public dispute of which all his clients would have been aware.

In essence, Price, representing sensibility, wanted landscape gardening to share with Italian painting the capacity to shock, to move emotionally, to override rational inquiry. Repton, forced to represent sense, summarised his argument in his red book for Attingham (March 1798) with an attack on "the futility of Mr Price's idea, that all improvements of scenery should be derived from the Works of the great painters."

Price and Knight had chosen a turbulent time for such a controversy. If the French Revolution was grounded in pure reason, it was also a revolution, and their Picturesque theory, despite its appeal to sensibility, was immediately associated with it. Theirs was "the Jacobinism of taste." In *Sense and Sensibility*, published in 1811, Jane Austen made light of the differences between the sensibility of Marianne Dashwood and the sense of her sister Elinor, but these two ideas dominated philosophical and political thinking, and Repton's own practice was to oscillate between them.

Meanwhile, Repton was still on the rise. The three landscapes discussed here parallel the three in the previous section: Sarsden for Moggerhanger, Blaise for Mulgrave, and Rûg for Courteenhall. However, in each case, the design has increased in intricacy, and by 1796, Repton had developed the model of his ideal landscape. This would have a small park—the smaller the better, so long as the surrounding arable land was hidden from it. In one place, it would have—wherever the topography most suited it—an "enchanted ground," ideally 40 to 60 acres close to primal nature, in which to indulge in Rousseauian reverie. Beyond, there would be extensive rides through the most pleasing parts of the neighbouring countryside, with villages smartened up and planted with flowers, and pleasing cottages (neither Gothic nor neoclassical but in their own vernacular style). The most important components would be the drives, approaches, and rides (what you could see from them and how these things were to be seen). Finally, the cottage and its garden, eventually to be among the most far-reaching and lasting of his achievements, were already taking shape in his mind.

opposite: On top of the hill at Rûg, embedded in the woodland, Fron Cottage—lodge, keeper's house, and Gothic folly all in one—appropriated the landscape

Cottages had been associated by the painter Thomas Gainsborough with moral innocence, and when dilapidated, a connoisseur like John Thomas Smith could indulge his fancy with their Picturesque condition "the patched plaster, of various tints and discolourations ... the mutilated chimney top—the fissures and crevices of the inclining wall—the roof of various angles and inclinations ... the decayed bee-hive and the broken basket—the fragment of a chair or bench—the slatternry of tubs and dishes scattered about the door—the mischievous pranks of ragged children—the intrusion of pigs—and the unrepaired accidents of wind and rain."[1]

However, such delapidations were not an invention on Smith's part. In 1775, Repton's friend Nathaniel Kent had lamented that "half the poor of this kingdom" were obliged to live in "shattered hovels." He went on to recommend an unaffected "natural" cottage style. Each cottage should have at least half an acre of garden to go with it: "I am far from wishing to see the cottage improved, or augmented so as to make it fine, or expensive; no matter how plain it is, provided it be tight and convenient."

Pragmatic and philanthropic, it is no surprise to find that Nathaniel Kent had been a mentor to Repton since they met at Felbrigg in Norfolk in the 1780s. He also worked for Uvedale Price at Foxley. Indeed, he may have introduced Price and Knight to Repton; and his was the vernacular "natural" style that Repton was to adopt for his cottages.

right: These oaks at Rûg are planted at the close intervals that are often to be found in Repton's parkland.

RÛG

— DENBIGHSHIRE —

CHAPTER EIGHT

RÛG

Repton's first visit July 29, 1794,
red book March 1795

THE DOMAIN AT RÛG falls from the north to "the small valley of Rûg," then briefly rises again to the ridge that carried the old Holyhead road across the south horizon. Beyond the road, the ground falls again to the "extensive vale thro' which the Dee takes its course," and beyond the river to the south, the first hills rise to 1,180 feet and on up to the upland moors of the Berwyn Range, overshadowing everything. Within the domain at its foot there were two buildings of merit: the rambling old manor house in the Rûg Valley, overlooking some fishponds, and a chapel built by the family in 1637 on the south side of the old road.

When Repton arrived in 1794, the landscape had little to offer beyond the grandeur of this setting, but he would have been comfortable with the family. Colonel Edward Williams Vaughan Salesbury, who had inherited Rûg in 1780, at the age of eleven, was distantly related to Lady Salesbury for whom Repton had worked at Brondesbury, and the family were loyal supporters of William Pitt the Younger.

Edward was just twenty-five in 1794, and he had the funds and energy for great works. But a job that Repton might have expected to find both rewarding and pleasurable was threatened at the outset by Price and Knight with their two publications on the Picturesque. The Rûg red book contains Repton's raw response to their attack, and his design returned to the Brownian style of his commission at Courteenhall. In political terms, since Price and Knight were Foxite Whigs, who looked favourably on revolutionary France, it was a decisive adoption of the Tory values of his client and of William Pitt.

The Salesbury family, though well-connected, was faux-ancient, having adopted the landed name in order to come into what was a substantial property, but Repton took what opportunity there was to offer his "assistance to preserve with pious veneration every vestige of Ancient or Hereditary dignity" as a means of showing his hostility to France ("At a period when the Ancient Family honours of a neighbouring country are rooted out with savage barbarity").

He therefore sought to retain one part of the old manor house and, though he did not mention it in his red book, he did save the ancient fortified mound or "motte" at the turn in the lake.

Repton may have anticipated criticism of his proposals from Price, Knight, and their friends when he wrote that "The outline of the plan, being thus laid down on paper, you will be able with more ease to explain the whole design to those friends whom every Gentleman wishes to consult on important occasions,

overleaf: The two-storey bay on front of the house provided views up and down the lake.
opposite: An array of doors and portico columns might frame a picture, but Repton's landscape first comes into view from the threshold itself.

and without whose approbation and concurrence, a plan is seldom executed with full satisfaction." He had used exactly the same words at Little Green in October 1793, and his client there, Thomas Peckham Phipps, was the cousin of Samuel Phipps of Ferney, friend and neighbour to Price and Knight.

Repton opened his attack by reducing their Picturesque argument to a statement that successful landscape gardening must depend on a knowledge of landscape painting. This premise he dismissed in a series of blows: a painting necessarily has a narrower field of vision than the landscape which is its subject; the wildness beloved by painters is not necessarily appropriate to landscape gardening which "may appear too artificial for a painter's Landscape"; and, finally, while they may have criticised him at Ferney for offering scenery without any object to focus on, he could deploy their own argument against them at Rûg where there were "that prodigious number of distinct objects" of interest "which no painting can represent, but which the eye delights in peering out and discovering in a populous district."

He then carefully distinguished "scenery" from a "composition" suitable for painting in the description of his sketches numbers five and six: "... a correct portrait of the scenery in front of the proposed house ... can hardly be deemed altogether such a composition as a painter would select." This insight marks a considerable step forward in sophistication from that of Ferney. He even exploited the idea of the "preparatory effect" that he had introduced at Courteenhall, comparing it to a view uninterrupted by foreground trees to carry the argument: "I doubt whether any one but a painter would stop short at this place, and not proceed beyond the trees, till he could at once enjoy the full display of this gorgeous and delightful valley."

Once Price and Knight were answered, Repton could be ambitious with the scale of the landscape that he was taking in. He had encouraged his clients

above: Repton made the best illustration that we have of the old manor house, shortly before it was taken down.
below: The new house was built further from the water, to a design sketched by Repton. The woods immediately behind it are his.

No. IV.
22
27
25
10
8
No. 6
N.B. The No. correspond with those on the Map from which view is taken.

Rúg's terraces "raise the earth in front of the building," and the view directly out is blocked by trees and the mounded earth beyond the lake.

to take up and improve their villages at Babworth, at Moggerhanger, and at Mulgrave. Here, at Rûg, he was explicit in recommending ridings out beyond the domain and a plantation of trees to incorporate Corwen, a town more than a mile away on the other side of the River Dee: "The prospect ... looks upon the town Corwen ... I have supposed a wood at the back of the town"), while in the other direction, he proposed to "proceed a little farther, and crossing the bye-road to Ruthin, look into that extensive vale thro' which the Dee takes its course from Bala to Llangothlin."

Views from the house remained problematic, however, as he was perfectly aware: "... the field of vision is much longer than any good picture ought to include; yet I should not do justice to my employer, if I were to represent this by two distinct landscapes, merely because they would then more closely resemble the works of great painters, while in reality the whole breadth is at once visible to the spectator." Both fronts offered open views of broad uphill slopes, with only the fishponds as foreground to the south. At Courteenhall, Blaise, and Stanage, Repton solved this problem by turning the front of the house into the side of the slope, but at Rûg the site and orientation of the house seem to have been determined before he arrived. His first proposal for a house in the Gothic style would not have helped, but pencilled into the red book are sketches of the neoclassical house that was actually built, its design attributed to Joseph Bromfield. These sketches show a bay in the centre of the south front and a large portico on the north. Both were architectural solutions to the problem of the panorama: the bay broke the view into three, allowing Repton to block the view from the house directly across the lake and favour instead the oblique views with longer reaches up and down the lake, which was to replace the fishponds.

To the north, the landscape was broken up in equally geometrical fashion, by means of the pillars of the

above: Repton's proposal for the view from the new house would have retained part of the old manor, but blocked the view in front in favour of a glimpse of the River Dee in the distance beyond.
opposite below: The village of Corwen conveys the scale of the hills behind it.

massive portico—a Brownian solution, and one that Repton would go on to propose at Moggerhanger in 1798. There was already woodland across the rising ground—a mixture of wood pasture and the established woodland of Coed-y-Fron, which acted as a belt enclosing the domain. Repton now added a "connecting" wood off the north front. This was planted over a quarry, a quarter of a mile from the house, but it appeared in views from the house to tie into the woodland of Coed-y-Fron. By scattering oak trees below it, Repton gave the appearance of woodland breaking across the contour and running down the hill, just as he had proposed at Mulgrave Castle. In short, he was working to give the hill a more dramatic and more intricate topography—he wished to confine the parkland to the south side of the house "within the small compass described on the map, which I suppose will not exceed forty or fifty acres," while making it appear to flow away indefinitely over the hill to the north.

This planting stands at "point blank," like the walnut at Courteenhall, and the various pre-existing field earthworks have been carefully smoothed away on both sides, wherever they would have been in view of the house. The thoroughly Reptonian Fron Cottage, though not mentioned by him, floats above his trees directly in front of the house, where the north approach runs up to the crest of the hill. As at Ferney, Moggerhanger, and Courteenhall, once beyond the parkland, this approach ran in a straight line over the arable fields.

Repton had justified this axial geometry in 1793 in his red book for Lathom: "Where the eye is meant

previous pages: The lake at Rûg, with its smooth and lazy sinuosity, is defiantly in the style that Capability Brown made famous.
above: The plain exterior of the chapel belies its extraordinary decorated mediaeval interior.

to survey the whole design at a single glance, it is assisted in its office by making one-half the counterpart of the other." Symmetry and unity in short can help the mind to take in a design. It will reappear at Blaise and Sarsden, but soon after Repton began his association with Nash it was to disappear from his landscapes: he would adopt geometry where it already existed, but otherwise it was confined to his gardens.

Repton was critical of the "lofty Portico of Greece," but besides dividing the view, it enabled him to move the entrance to the north front, thus reserving the views south over his lake for the bay window of the drawing room. However, his approaches would still have to run in from the old Holyhead road on that south side. He therefore proposed one route past a covered seat at the top of the park and through his wooded pleasure ground, and a second from the west to run through the equally deep plantation of Stable Wood (both routes were to be altered when the old Holyhead road was closed). Both approaches would show the house to disadvantage from above, however, and Repton's solution to this was to "raise the earth in front of the building" and hide the cellars "to give a good fall of ground from the house.", as he had done at Welbeck and was to do again at Pentillie. This would emphasise the horizontals rather than the height of the building, "and thus the front will appear to stand on a gentle eminence."

There is much that is Brownian in Rûg's design: the surrounding woods, the extent of grassland, the reliance on geometry, and the ridings, but the most obvious detail is the river-like lake that winds across the front of the house—constructed at a time when, at Babworth and Mulgrave, Repton had turned towards the "prattle of the brooks." He may have been badly hit by Price and Knight, but at Rûg, so close to their territory on the Welsh Marches, he was not going to show it.

above: The gates at Rûg, though shut firmly against the Picturesque, embrace the antiquity of the Tudor rose. overleaf: Oaks in the parkland at Rûg.

BLAISE
— BRISTOL —

CHAPTER NINE

BLAISE

Repton's first visit August 1795,
red book February 1796

"Blaize Castle!" cried Catherine.
"What is that?"
"The finest place in England—worth going fifty miles at any time to see."
"What, is it really a castle, an old castle?"
"The oldest in the kingdom."
"But is it like what one reads of?"
"Exactly—the very same."
"But now really—are there towers and long galleries?"
"By dozens."

The dreadful John Thorpe in Jane Austen's *Northanger Abbey* is trying to persuade Catharine Morland, a neophyte of the Picturesque and of Gothick Romance, to give up the walk she had arranged with the thoroughly rational Tilneys in favour of a sensation-seeking carriage drive to Blaise Castle.

There had been already been gardens at Blaise for more than one hundred years. However, the subject of Austen's delighted satire was the creation of Thomas Farr. He had made a fortune trading in sugar and bought 110 acres and the old manor house from the Henbury estate in 1762. His land, backing onto the village of Henbury, rose up to Blaise Hill and took in the extraordinary gorge of Hazel Brook.

Farr was a friend to both the statesman and aesthete Edmund Burke and Valentine Morris, who had plantations in the West Indies and was making his famously Picturesque garden on the cliffs above the River Severn at Piercefield. Perhaps Farr aimed to outshine them both at Blaise, for its cliffs and deep chasms were a gift to the Picturesque. His centrepiece was the castle itself—a toylike eye-catcher on top of the hill, with a dining room on the first floor and views of the shipping on the Rivers Severn and Avon. This castle he defended with look-out bastions, wooden cannon, and terrible legends: a Robber's Cave, a Lover's Leap, a root-house, and Giant Goram and his bath house, known as the Giant's Soap-Dish. Repton made it sound more like a mining disaster: Farr had had the "bad taste" to propose an "artificial river ... about the width of a common navigation canal, secured by different heads or dams [with] all the horrors of fire and steam, and the clangour of iron chains and forcing pumps."

To be fair, Farr had encouraged the public to walk in, and it may have been their "barbarous taste" that he sought to gratify. However, his ships were blockaded in the American Revolutionary War, he was bankrupted in 1778, and the property was eventually acquired in 1789, within a week of the fall of the

previous pages: Blaise Castle was a prospect tower, built by the merchant Thomas Farr in 1766.
opposite: Repton placed William Paty's villa on the only appropriately smooth ground on the estate, with a view to Blaise Castle at right angles to its south-west front.

Bastille, by Repton's client, the banker and Quaker, John Scandrett Harford.

Harford immediately annoyed his neighbours by cutting down trees. Repton weighed in to support him, while acknowledging that it might be unpopular: "... tho' fully aware of the common objection to cutting down trees, yet it is only by a bold use of [the axe] that the wonders of Blaise Castle can be properly displayed."

Notwithstanding this felling, Repton aimed to make the place more private and less sensational, for Harford's Blaise, like Moggerhanger, was to become a "retreat from the bustle of the world ... perfectly secluded from the 'busy hum of man.'" To that end, Harford had bought more land to the south and west, and Repton designed a long principal approach across it that would avoid Henbury village. This may have been as unnecessary and extravagant as it was short-lived. It effectively divided the landscape into three (the Royals, Coombe Hill, and Home Park), as did the approach at Sarsden, but Repton could make of it a Picaresque narrative, taking the traveller through a "numberless" series of "beauties which may be brought before the eye in succession by the windings of a road." So, it began with a Gothic lodge and then proceeded through woodland with occasional

previous pages: Repton's challenge was to find a route by which his new approach could cross the dramatic gorge of Hazel Brook.

views under the branches of beech trees, as at Babworth and Honing, to the meadowland of the Royals, until he could offer a sudden preparatory glimpse of the house through a break in his shrubberies. Then came a thrilling series of bends; he boasted of their construction in his red book for Burley-on-the-Hill (1796): "At Blaise Castle … I have lately finished one of the finest approaches in the kingdom; altho' for many hundred yards the sides of the hills were so steep, that I had to be let down by ropes to mark the line of the road." Such acrobatics were necessary if he were to traverse the gorge of Hazel Brook. The switchback would lend excitement to fast drivers like Jane Austen's Willoughby, thinking only of curricles and hunters, but as he had recommended at Moggerhanger, his approach would also be convenient for "the aged and the infirm … shewing this situation … from the windows of a carriage with all the interest of surprize and novelty." Therefore, if it was to remain "in strict character with the wildness of the scenery, and excite admiration," it was also to avoid "any mixture of that terror which tho' it partakes of the sublime, is very apt to destroy the delights of romantic scenery."

He also had to "appropriate" the additional acreage and so proposed a Woodman's Cottage, as he would

opposite: The beech trees suppress the understorey and make it possible to look deep into the woods.
above: In the woods planted by Repton, only the most dramatic cliffs show through.

top: Repton's panorama shows a pastoral setting for the new house.
above: The Robber's Cave gave a view to the down at King's Weston.
right: Repton's plan shows the winding route of his approach from the south

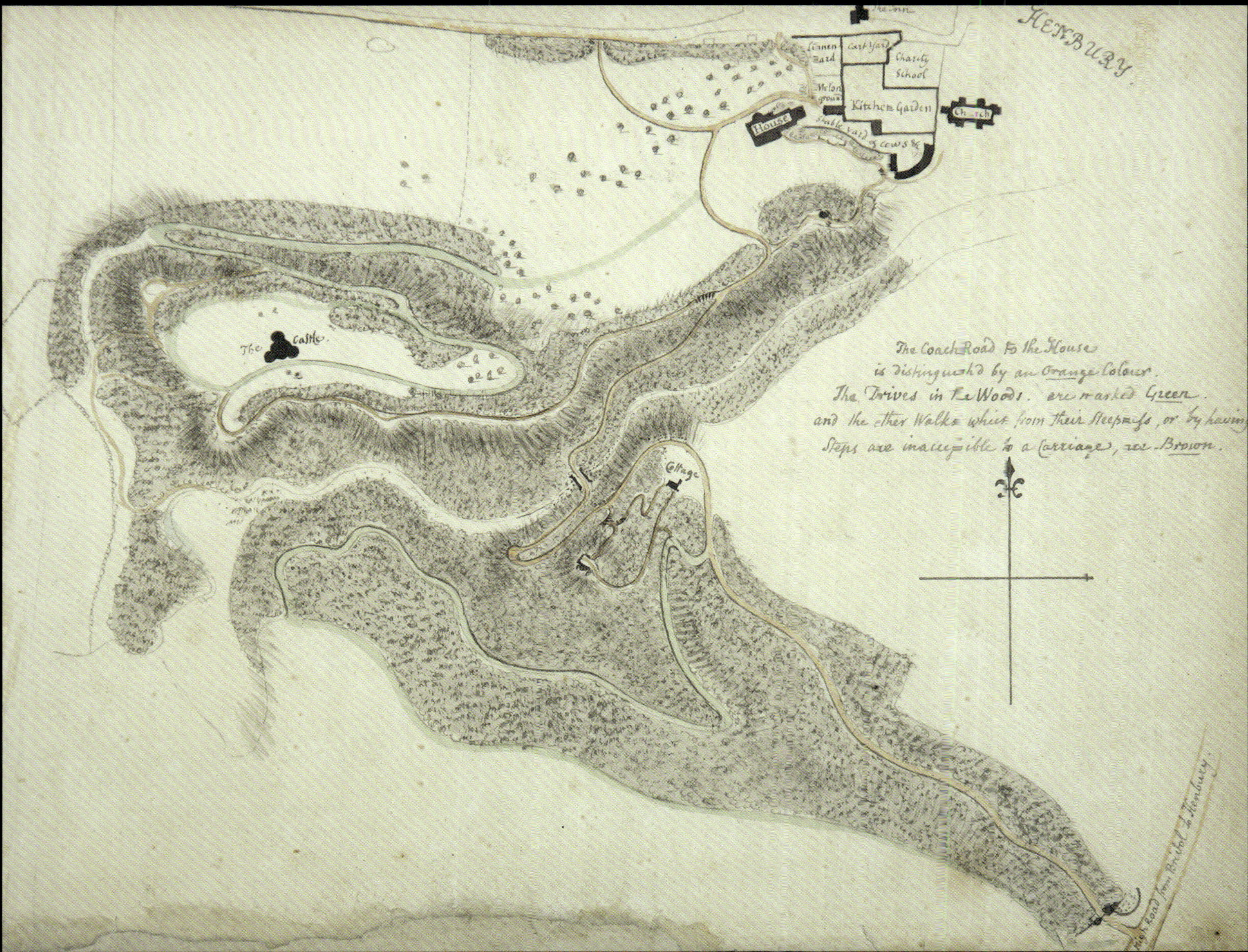
HENBURY.
Cart Yard
Charity School
Melon ground
Kitchen Garden
House
Church
Stable Yard for cows &c
The Castle
The Coach Road to the House
is distinguish'd by an Orange Colour.
The Drives in the Woods. are marked Green.
and the other Walks which from their Steepness, or by having
Steps are inaccessible to a Carriage, are Brown.

at Moggerhanger in 1798, which, like the covered seat at Rûg, would front the woodland and "seem to belong to the proprietor of the mansion and the castle, without affecting to imitate the character of either." Here Repton made his first attempt to define his vernacular cottage style: "It must look like what it is, the habitation of a labourer who has the care of the adjoining woods; but its simplicity should be the effect of Art and not of accident."

He was to conceal the valley floors (as he had at Ferney and Mulgrave) and the garden terraces on the side of the gorge southeast of the house (as he had at Ferney). This time, he advised that "all the yawning chasms be hid by plantations, rather than let any traces remain of works, done under the influence of such barbarous taste."

Nonetheless, several themes of the red book illustrate his growing Picturesque sensibility. He rejected the proposal that the new house should be built on top of Blaise Hill because it is so difficult to make a landscape out of a prospect. He chose instead a site close to the old manor house in one corner of Harford's estate. At Port Eliot in 1793, he had already denounced as false taste the rule that "the house should stand nearly, in the centre of its park, and detached from every surrounding object," and this location gave him greater length for his principal approach. Furthermore, the Bristol architect William Paty (1758–1800) had already designed the house in neoclassical style, and Repton could best accommodate this on a site where he could separate its polite form from the Gothic and sublime imagery prompted by the gorge. In fact, although the Gothic Lodge broke Repton's rule, published in 1803, that the architecture of entrance arches "should correspond with that of the house, in style, if not in order; that is, the Grecian and Gothic should be kept separate," Blaise marks an early, though reluctant, acceptance of their coexistence: "The stranger will be agreeably surprised to find that on quitting this wood, he is not going to a mouldering castle whose ruined turrets threaten destruction, and revive the horrors of feudal strife; but to a mansion of elegance, cheerfulness, and hospitality where the comfort of neatness is blended with the rude features of nature."

He also advised on the orientation of the house, with Blaise Castle at point blank (like Fron Cottage at Rûg) off its southwest front, while the principal front, to the southeast, had a view that would have pleased Gainsborough, with the Woodman's Cottage on the right-hand side and the land falling away across the lower ground of the Royals towards the east.

There are close parallels with Mulgrave. In both cases, Repton came into a landscape already undergoing development in the Picturesque style. In both he completed the work; in both there were deep

above: Repton's before-and-after pictures show the rambling manor house replaced by a trim box.
opposite: Blaise Hamlet is an attempt to capture the lost innocence of olde England, as recalled in the early nineteenth century. The Dairy (below) was built in 1803 next to the house, a forerunner of Blaise Hamlet.

romantic valleys and views to open water and shipping; both had neoclassical boxlike houses set back from this scenery; in both he created an "enchanted ground" with seats, caves, and waterfalls; and in both his design was characterised by numerous long, looping drives and cottages in his vernacular style. Since his clients, Harford and Lord Mulgrave, had very different backgrounds, it would seem that Repton designed primarily to suit the land rather than their taste. His skill lay in persuading them to enjoy the particular character of their territory.

An account of Blaise would not be complete without mention of Blaise Hamlet. This was not built until 1810, and is generally attributed to Repton's partner John Nash.[1] However, the evidence favours Repton's son George Stanley, then twenty-four, for the detail and much of the design of these extraordinary cottages. Their style was a whimsical extension of vernacular forms already employed by Repton's old friend and colleague the Norfolk architect William Wilkins senior. Repton remained involved with Blaise and had recommended and designed cottages ornés as early as 1791, when the traveller John Byng commented on the thatched rectory at Scrivelsby as "repair'd upon Mr Repton's plan." He had described the village at Babworth as a "new created scene" and had proposed a model village for High Legh, thirteen years before the building of Blaise Hamlet. His client there, George Legh, reminded him to "make the Green of High Legh a village ... surely this was the idea that first struck you as mentioned in yr [red] Book."[2]

Repton amplified this proposal in a note to Legh after a visit with John Nash in 1797: "Both Mr Nash & I shrink from the idea of a row like almshouses—or the appendages of a manufactory ... especially as we should like to make the doors open on the sides or back of the houses that the fronts to the gardens & Turnpike might be kept Neat." This might be a blue-print for Blaise Hamlet, even to the side door and neat front garden, which Repton popularised. It was to be fenced off and maintained by the cottager, but would contribute more to the pleasing character of the community than to the pleasure of the individual. Indeed, he may have invented the front garden at High Legh, for the contrast with the village at Courteenhall is marked.

Blaise Hamlet was acclaimed, as the German Prince Pückler-Muskau wrote home after his visit there in 1828.

> *The proprietor, Mr Harford, has sought to realise the ideal of a small village. A beautiful green space in the midst of the wood is ringed by a meandering path with nine dwellings distributed along it, all of different design and diverse materials: one is sheathed in fieldstone, another in ashlar; this one in brick. That one in wood; one is roofed with thatch, another with shingles, the next with slate, and so on. Different trees have been planted around each cottage, and each is wreathed with various types of clematis, climbing roses, honeysuckles, and vines. The dwellings are separate, yet together form a whole. Each has its own garden, and they share a common well, which stands in the middle of the green and is shaded by several groups of old trees. The gardens, bordered by pretty little fences, create a pretty garland of vegetables and flowers around the village. But what crowns the entire settlement is that all the families here are poor families, and the generous owner allows them to live in the houses free of charge.*[3]

Despite its success however, these cottages had little lasting impact on Repton's work. Blaise Hamlet was an "affected" *jeu d'esprit*. The simpler vernacular Woodman's Cottage was more influential; Repton returned to it in 1803, adding to his account: "Its form is meant to be humble, without meanness." The cottage was to express the woodman's station in life, without demeaning him. It was, in James Malton's terms, "natural" (making use of "those materials which the spot whereon it is to be situated produces, and often of the cheapest kind, rude and unadorned"), whereas the hamlet cottages were "affected" ("the returned East Indiaman may be deemed absolutely necessary, as having brought a principal part of the required materials").[4]

previous pages: The nine cottages at Blaise Hamlet are each architecturally distinct and all are remarkable.
opposite: After a return visit, Repton was mortified that "from the obstinacy and bad taste of the Bristol mason" the Gothic entrance to his south approach was "built of a dark blue stone, with dressings of white Bath stone."

SARSDEN

— *OXFORDSHIRE* —

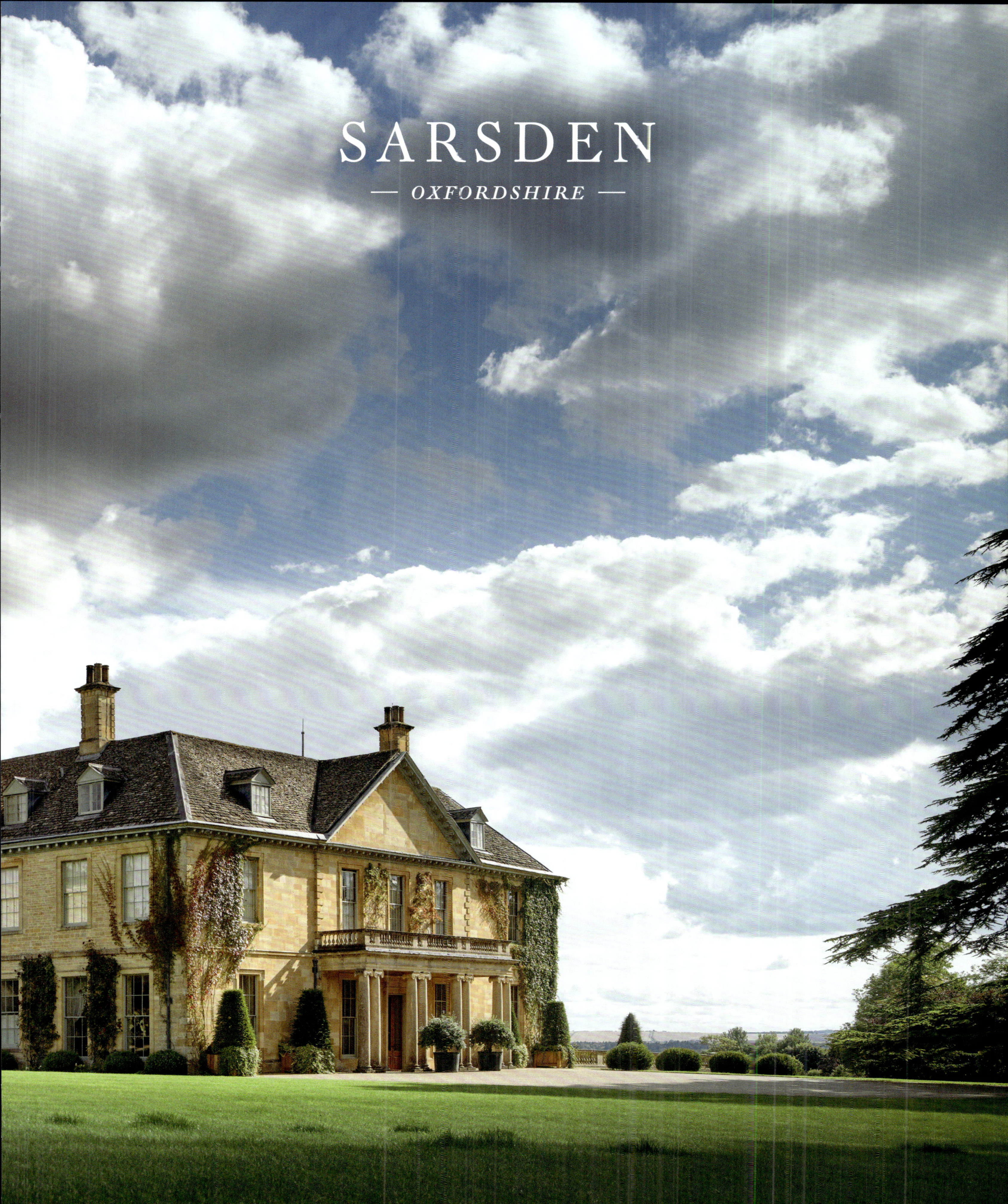

CHAPTER TEN

SARSDEN

Repton's first visit September 20, 1795,
red book March 1796

Sarsden is one of many landscapes along the old road from Oxford to Stratford. The ground is gently corrugated, with the house on the crest of one wave, the village of Churchill on the next to the north, and the valley of the Sars Brook between them. To the south and southeast, the land falls again from the house to its pleasure ground before rising to Fairgreen Farm; but Sarsden House sits on a broad hill, and these corrugations and valley floors are out of sight from its windows. The network of lanes that feed into the road would have further challenged any owner to realise a single unified composition. Nonetheless, two sets of great stone gate piers on the Chipping Norton road and at Fairgreen Farm indicate just that ambition on the part of the Rolle family, who had acquired Sarsden by marriage in 1731.

In 1792, however, before any such 1,000-acre park could be enclosed, James Haughton Langston, a spectacularly wealthy wine merchant and, like Godfrey Thornton of Moggerhanger, a director of the Bank of England, bought the estate. When he died in 1795, his son John (1758–1812), a Pittite politician, commissioned Repton, who announced that rather than make a great park, he would concentrate on the valley to the south which "furnishes ample store of all that the most fastidious improver can require; here is wood and water and undulating forms of ground, with just so much of distant prospect as may interest the eye without fatiguing it." He had first to acknowledge, however, that the view from the Gallery would be difficult to paint (repeating the argument he had made at Moggerhanger), "because it consists of wood and lawn without any leading object to fix the attention." His solution was to make a lake, but to plant shrubs and Lebanon cedars at point blank and to construct objects on each side of them—the Bridge, rusticated and "of three small arches rather than a single large one; because it will make the water appear of more importance," and the Temple Boathouse.

Sarsden shares with Moggerhanger Repton's preoccupation with peep-views: from the house to the Bridge ("I have ventured to mutilate a large tree near the house, to shew the bridge under its branches"); from the approach ("betwixt the group of trees at G, and the sunk fence at H, the valley, with all its wood and water presents itself, to the greatest possible advantage"); and from his "line of communication" between house and lake to the Temple Boathouse. Repton regarded this walk as "one of the greatest improvements both in point of beauty and convenience, that I have the honour to suggest." It was not possible to keep the grassland around the house free of farm

previous pages: The entrance is on Sarsden's shortest front—the north. The chapel is on the east side of the house.
opposite: Someone (surely Repton) pencilled the dimensions of the rustic Temple Boathouse into the red book.

stock while the family was in residence, and those who wished to walk to the water safely would need the protection of fences. Repton's pleasure grounds at Holkham, Cobham, and Rûg all served a similar purpose, but this walk was bound to cut through the middle of the parkland and so presented a greater challenge: "It is not very easy to protect a walk across a lawn, by a double fence without giving it an air of restraint and confinement; but by varying the kind of fence, and by sometimes hiding it entirely, the walk will be secured from cattle on both sides, without exposing much of the fence, and also without excluding entirely views of the water from the new approach." Repton was forced to abandon any idea of making a single coherent Brownian domain, but compensated for this through a rich variety of planting along the walk. So, at Sarsden, one might imagine a parterre between the wings of the house, as he had recommended at Cobham; then a couple of copper beeches to break the view and provide the painterly *repoussoir* effect that he was to bring off at Moggerhanger; and below them a flower garden, tightly enclosed from view by a substantial bank planted with shrubs, such as he had proposed at Courteenhall. After that, an area of intricate or mazy shrubbery—small groups of shrubs and trees that he had described in his red book for Purley in 1793: "shrubberies, flowers, and walks ... very ill described by a map, and still less capable of being represented in landscape," falling to a loose grove of cedars at point blank, and finally opening to the lake, which Repton formed out of a series of fishponds. The new water had a walk around

previous pages: The bridge was designed both to be an object in the view and for traffic on the principal approach.
opposite: These cedars of Lebanon blocked the view of the house from the water, making it a private retreat.
above: The plan in Repton's red book has the approach that is still in use today pencilled onto it.

Repton's revisions to his Sarsden red book were extensive and his cedars were to block this view.

it, ornamented with flowering shrubs. Next to it was an orchard and Keeper's House, like Circular Cottage at Blaise, set in its own small landscape. This area was another "enchanted ground" for Repton, not sublime like the quarry at Ferney Hall and the deep gorges at Blaise and Mulgrave, but a place of Arcadian retirement that remains a delightful jewel, threaded onto the sequence of effects that are to be had from the approaches.

Repton wrote so much about approaches at Sarsden because, as he made explicit himself, he had come to place increasing importance on the routes within and beyond the domain: "There are many objects of future improvement which do not fall within the compass of this map, such are the great wood thro' which a drive might pass to the tower above Churchill, and also the manner in which the distant hills might best be planted, both with a view to ornament and profit." He had made similar claims for the virtue of long ridings through the countryside at Cobham, Honing, Moggerhanger, and Rûg, but now he had ideas for improving this countryside, not just with a "view to ornament" but for profit. Thus, he dismissed the Picturesque "romantic forest scene of unprofitable thorns and briers ... as it is the appearance rather than the reality of extent that is particularly necessary, and as there may be some sacrifice of profit, if not of convenience, in converting ploughed fields into grass land; I always endeavour to make the park as small as possible." By such means, he hoped to "contribute to the good of my country." Corn was needed and prices were high—the war gave Repton an additional reason for restricting the area of parkland. Hence, the advice he had given at Rûg, he gave again: to hide the fences, conceal the "corn lands," and so carry "the appearance of park ... quite over the brow beyond these large tracts of wood to an unlimited extent." Indeed, his comment that "a wood over which a naked horison is visible, loses half its importance, unless that horison be bounded with such vast hills as deserve the name of mountains" reads as though he had Rûg in his mind. His parkland was both to look more extensive than it really was and to blend into the surrounding countryside.

right: Repton's second choice for an approach leaves the valley floor on an easy gradient shaded by beech trees.

Work stopped when John Langston died in 1812, and resumed when his son, a second James Haughton Langston (1796–1863), came of age in 1817. James Langston commissioned the Reptons to design a number of cottages, whose names indicate that they were built for Langston's employees (Green Lodge, Keeper's House, Drive Cottage, the Kennels, and so forth). They were spread over an estate that took in the neighbouring village ("a drive might pass [through the great wood] to the tower above Churchill"), an extensive acreage down the Sars Brook valley to the east, with the dower house (Sarsgrove) in "the great wood") and Sarsden Rectory. New Farm was given a new front and a modest front garden, as Repton had suggested, but all these buildings (except Keeper's House in the pleasure ground) were in the natural, rather than the affected, style.

There is no record of how involved Repton himself was with these cottages, but he was still working continually with his son George, and such collections do crop up again and again on his landscapes—Attingham, Laxton, Stoke Edith, Stoneleigh Abbey, and Woburn Abbey are just a few examples. The argument in the Sarsden red book is also telling: "As no part of the home premises can be rendered visible from the turnpike, it might perhaps be advisable to continue a respectable belt or skreen of plantation along the side of the road, and by moving the two great piers to the corner of the private lane an approach to some place would be more distinctly marked." The piers are three-quarters of a mile from the closest parkland, but they were "to attract attention, or to excite the idea of greater beauties." As a "preparatory effect," they extended the landscaped domain. This idea was on Repton's mind: at Wentworth Woodhouse in 1794, he was to discuss "this great plan for improving the ... Environs of Wentworth-park." So far as Richard Payne Knight was concerned, Repton had transgressed when he suggested for Tatton Park in 1792 that "even a mere stone with distances," which "may bear the arms of the family," would by such a means "impress the mind with a sense of its influence." Repton use of "mere" here reads like a conscious riposte:

above: Repton proposed moving the "two vast piers" to mark the beginning of the approach to the house.
opposite above: At Sarsden, Repton recognised the charm of a more intricate view.
opposite below: The lodge to Sarsden Glebe is a paradigm of the "natural" style, executed by Repton's son George.

"The extent of property belonging to the house, confers no importance on the place unless distinctly marked by some appropriate objects or by a character different from that of a mere farm."

Central though ridings and approaches might be to his developing idea of landscape, Repton faced some difficulty in finding a good line at Sarsden. As at Cobham, he criticised the existing approach because "Altho' the house stands actually upon a considerable eminence, the approach is made to descend towards it." His first choice ran in from the east down the Kennels Belt along the floor of a gentle valley. This provided good access from the A361 and allowed him to take a straight run of road across the arable land between the piers and the parkland proper, as he had at Courteenhall, Moggerhanger, and Rûg. However, it left him with the difficulty of finding an easy ascent from the valley floor with a good view of the house on the way. His solution was to take his approach through the pleasure ground itself (as he had at Rûg), along the south side of the lake, over his bridge, and up to the house through the west park. Since the last section must have been grassed, this could only have been used in summer, like his Northampton approach at Courteenhall, but he excused himself with the comment that "It may perhaps be asked by some, who only look at the lines on paper, why the approach does not proceed from the farm to the mansion in one bold curve, instead of the waving line there marked; it is, because the whole course of such a line would have been uninteresting, as the eye would have been directed from the chief beauty of the place." The version that he finally settled on, pencilled on his plan but described in his text, ran on the north side of the valley, set at the back of a broad shelf of land so as to be invisible. This had glimpses of the house "yet there is enough of it seen to mark its importance," with "the combination of the chapel."

In his red book, Repton confined himself "to objects nearer home, as more essential to be first compleated." Yet we know that he returned and revised (there are pencilled additions; the map is out of place; the plan and elevation for his proposed lodge, never built, has been inserted), and the landscape that he left actually consisted of a series of threadlike developments, strung out along his approaches and drives, which set a smaller area of parkland (220 acres) within a landscapable domain similar in area to the 1,000 acres apparently intended for a park by the Rolle family when they sold the estate to the Langstons.

opposite: George Repton designed Sarsden Glebe in 1818. No buildings at Sarsden went untouched by the Reptons.
above: George Repton also designed the magnificent church tower at Churchill.

1795–1802

THE NASH YEARS

IN 1796, NAPOLEON Bonaparte, twenty-seven years old, took command of the French Army. In 1797, he bundled the Austrians out of Italy and forced a peace. With the loss of Austria, a British victory against France seemed inconceivable.[1]

Napoleon then turned on Britain. Declaring that "Truly to overthrow England we must occupy Egypt,"[2] he advanced into North Africa in May 1798. His fleet was annihilated by Admiral Nelson at the Battle of the Nile, stranding his army and cutting off its supply chain. Most of the European monarchies saw their opportunity and formed the "Second Coalition" to fight him. But it wasn't enough. In November 1799, having put down another rebellion in Paris, Napoleon seized dictatorial power as First Consul and offered a European peace. The coalition refused his terms, so he marched over the Alps and smashed the Austrian Army at Marengo in 1800. Thus ended the Second Coalition.

This conflict remained cultural as well as military, for Napoleon had given the French press its freedom in 1796, while William Pitt had suppressed radical political thought in Britain. Nonetheless, Pitt resolved to continue the war. By the end of 1797, he had proposed a combined tax to fund it, intended to target the wealthy. In March 1798, there came martial law, and when a second uprising in Ireland was crushed at the end of that year, his government pushed through the Act of Union (January 1, 1801), under which Ireland was directly ruled from Westminster. But persistent cold wet weather had devastated the harvest in 1799, and the bread price tripled. The demand for peace became overwhelming, and, in March 1801, Pitt resigned.

Pitt had opened a schism in British society. On his side, the Evangelicals and Tories were opposed to France and its Revolution, concerned to control radicals and the mob, and increasingly inclined to the church and the traditional subordination of lesser to greater (whether people or houses). They shunned conspicuous displays of wealth. On the other side were Pitt's great rivals, Charles James Fox, his Whigs, and the Prince of Wales—entertaining but contemptible.

By 1796, Repton's likely clients were threatened with tax and riot. He himself was under attack from Price and Knight. When all certainty was gone, rescue came from an unlikely source. The architect John Nash trained with Sir Robert Taylor, was bankrupted in 1783, moved back to Wales to rebuild his practice, and befriended Richard Payne Knight's cousin, Thomas Johnes, at Hafod. Repton first met Nash in 1791. In 1795, the latter became a favourite of the Picturesque Movement when he designed that famous triangular house on the front at Aberystwyth for Uvedale Price. Now they went into partnership. By working with him, Repton must have thought he could stifle any further attack from Price and Knight.

Nash was a spendthrift—slapdash, disloyal, self-serving, an untrustworthy friend, and a thief of other people's ideas. He was temperamentally and socially a Foxite—neither a purist, nor an antiquarian—but he had brio and charm. Their partnership should not have worked, but it lasted for about seven years. They joined forces at fourteen sites between 1796 and 1800. In 1802, Nash moved on, accepting the patronage of the Prince Regent, later George IV. Once appointed as architect to the Crown Lands, he reshaped the west end of London. Repton never forgave him for the break, but his partnership with Nash triggered major changes in his style. With a renewed confidence in his ability he began to explore new forms of architecture and design in landscape.

opposite: The summer house in the American Garden is a secret haven in the heart of Luscombe's parkland.

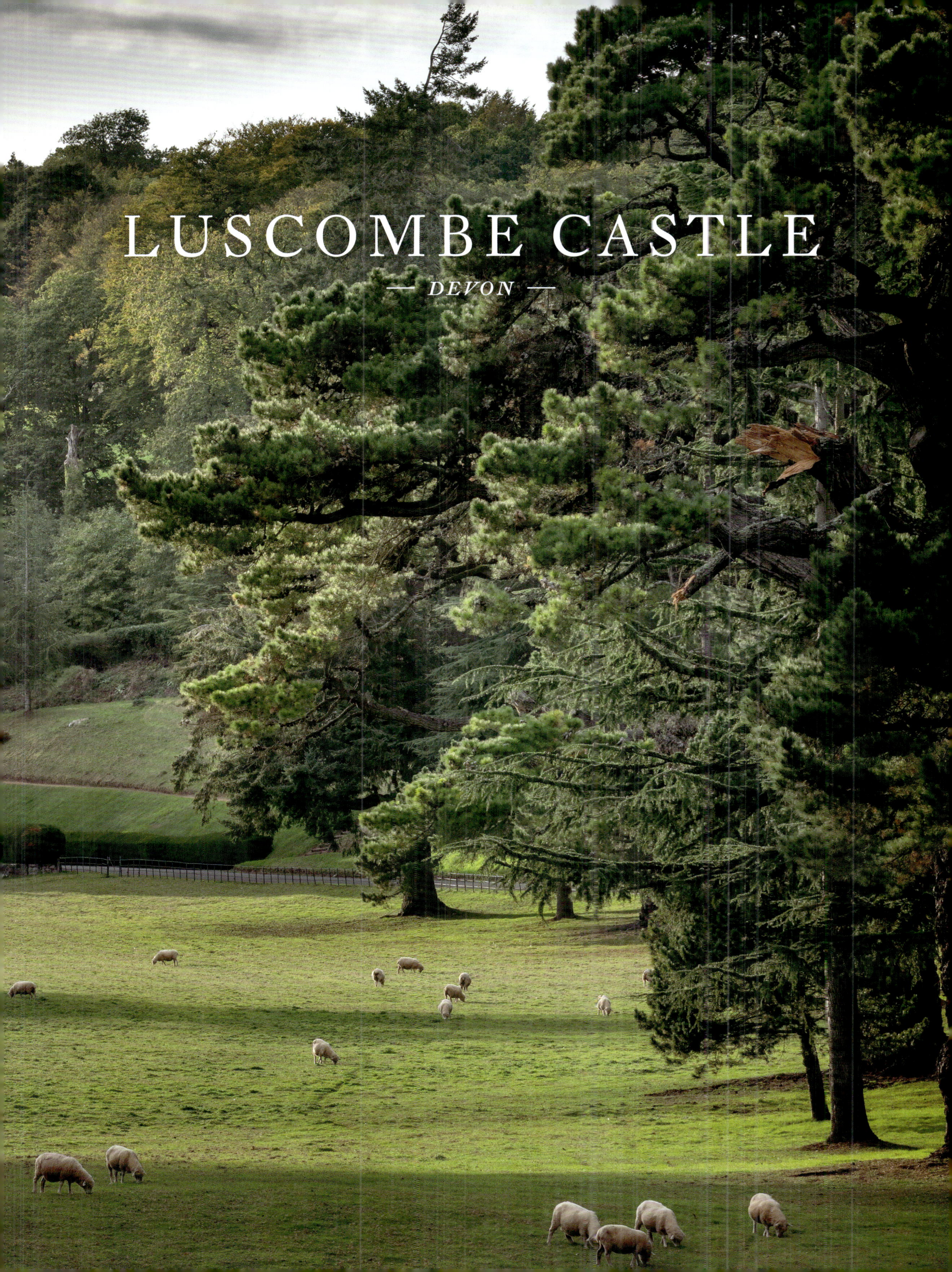
LUSCOMBE CASTLE
— DEVON —

CHAPTER ELEVEN

LUSCOMBE CASTLE

Repton's first visit February 1799,
red book July of the same year

C. Hoare & Co. was a well-established private bank that remains a family business today. Henry Hoare I (Charles' great grandfather) was able to buy Stourhead in 1717 and the bank weathered the bursting of the South Sea Bubble (1720) so well that Henry II ("The Magnificent"), inheriting Stourhead in 1725, was able to lavish money on his famous garden there. At the outbreak of the French Revolutionary Wars, the bank was run by another Henry who greatly, and with legendary discipline, increased the business. Charles Hoare (the "C" in the bank's name and Repton's client) had been made a partner in 1787. Hoare then set about purchasing land from the old farming community around Dawlish in order to make a place for his wife, Frances, who had poor health, within sight of the sea and the bracing marine air.

When Repton first visited, the Luscombe valley and its downs were predominantly small grass farms, woodland, and rough pasture on the steep slopes, with a small amount of arable on the valley floor, more on the flat hill tops, and moorland on the higher ground out to the west. Repton had not had many opportunities to build an estate from scratch with such a sympathetic owner, for Charles had the money, and Luscombe was a congenial site. It may still have been in several ownerships, but Repton was happy to advise on that land as he had Honing. He had a free hand to place the house, and he had the favoured architect of Price and Knight, John Nash, on side.

Indeed, it has been suggested that Luscombe Castle owes more to Nash than to Repton, but much of its design had been prefigured in earlier red books. Repton had recommended "Castle Gothic" as early as 1790 at Scrivelsby because "the stile of Architecture ... gives scope to the imagination which will be led to suppose it much larger and more magnificent than any modern building of twice its dimensions would appear in this Situation." What most attracted him to Gothic was its license for irregularity. Gothic could also solve the problem he had come across at Cobham, where he had found it a disadvantage to approach the house from above "because the roof and chimneys will for a long time present themselves over the rising ground." A neoclassical "Grecian" roof would be unsightly, but Castle Gothic could enrich the roof with Picturesque "turrets, battlements, corbles & lofty chimneys."

The siting of the castle is the overriding theme of the red book, and it seems to have been written to justify the choice of site *post facto*. This had been a radical decision—it might have been built on the higher ground to the west, so as to have a glorious

previous pages: The castle is beautifully proportioned to the valley in which it sits.
opposite: With their great windows, the library and the conservatory were designed to frame and set off the views over the Front Park to Dawlish.

To create his "burst" on the Church Approach, Repton carried the road uphill to slow the horses then turned it around the shoulder of the ground to show off the castle at the bend.

view down the full-length of the valley towards Dawlish and the sea; it might have been built on the floor of the valley, where the old Luscombe Farm had stood. Repton chose a place close to the valley floor, wherefore it would be "in appearance a large place, because the ground in every direction will be appropriated to the Castle" because "wherever turrets, or chimneys appear rising out of the wood, the Imagination will fill in the space between them with buildings, which it supposes the wood conceals." However, by moving it west of the farm, he could use the castle to divide the landscape into two parts: the Front Park running down from the house towards Dawlish, while the Back Lawn took in a landscape of quite a different kind, on the steeper, more dramatic and Picturesque ground to the west. This composition consisted of a relatively simple "skreen of plantation" on the north side of the valley with a drive running through it and views across the valley to the more dramatic folds to the south, emphasised by the scrolls of woodland that follow the shape of the ground. Repton was to repeat this effect at Ashridge and Woodchester, and his planting method ("it will be better to cloath the whole with wood, which may be opened & partially cleared hereafter, as the beautiful points of view will best direct") at Brightling.

The resulting parkland is a superb example of the English landscape style, recognised at the time as forming "the closest copy of natural scenery as yet produced by means of plantation, in any part of the kingdom." Essentially, it consisted of a band of woodland running up on each side of the valley, diversified with single trees and clumps (principally beech, often in association with holly), carefully placed between the plantations and the valley floor. Those little knots of planting used to break the edge of the plantation at Courteenhall are here Repton's central concern.

By placing the house on the valley floor Repton avoided the tussle (which he had had to address so often) between profitable arable farming and grass;

above: Repton planted Holm oaks as shrubs around the American Garden. They have grown since.
opposite: Repton composed his views to be seen from the ancient oaks that he preserved at Luscombe.

there would be no possibility of seeing the arable in the neighbouring Aller Valley. Yet, there was nothing tortured about this, as there had been at Sarsden, where he had been obliged to drop the approach to the floor to avoid views of arable and then had difficulty engineering the final climb up to the house. The principal approach—from Dawlish—had an easy gradient along the north side of the valley.

In his red book, Repton struggled to put his planting advice into words; he had managed a little more detail at Rûg and Sarsden. It was clear that he was doing something complicated. He did draw attention to one feature of special interest to him: "... the plantation to the South of the House, instead of cutting across the side of the hill in one line, will project forward on the brow of a second hill, & recess into the accidental valley, so in the more distant woods, the precise outline must be guided by the shape of the hill, taking advantage of any irregularity of ground." Repton seems to have liked this "accidental valley" well enough to make his entrance to the park, at Front Lodge, immediately opposite it. It fully justifies the praise that his work at Luscombe received. Essentially, this is a T-shaped clearing running south at right angles to the main valley along the course of a subsidiary fold in the ground, and then turning west on the higher ground. However, it is planted with beech in such a way that the ground appears now to flow into one arm of the T, and now into the other, with no appearance of effort or artifice. Beech

overleaf: Repton's planting of the "accidental valley" shows him to have been a master of parkland design, making a dramatic and ever-changing hollow out of what was little more than a topographical ripple.

is the best native tree for picking out and emphasising the form of the ground. In summer, it suppresses the understorey and, in the autumn, its fallen leaves cast rust-brown shadows across the grass.

Price and Knight would have had little to complain about here—not only the planting and the Castle itself, but also the modest size of the parkland and the pared architecture of the lodges, subordinated to the Castle, were all in keeping with Picturesque ideas. Yet, in its apparent simplicity, the overall character of the design is thoroughly Brownian. This shows itself both in the planting and in the views—for this is a landscape articulated not by peeps carefully framed to produce a single effect, but by more generous views, spaced around the landscape, that may sometimes show more than they need to, in order to form a perfect picture. His viewpoints consist of small mounds with a tree in the middle and a bench seat around the trunk, and small levelled bays beside the drives around the park, where a carriage might pull up to take in the scene, or a seat be set out for an afternoon picnic. These bays are a constant in his landscapes, even though never directly mentioned in his red books, but Luscombe is replete with them. They would have suited an invalid, needing exercise as much as she needed rest, but they also mark a change from Blaise, where Repton had lavished so much attention on the views from the approach itself.

The view down to the Castle from the northernmost part of Luscombe Wood is the finest today and was frequently painted in the nineteenth century. The south facade was another favoured aspect for artists, and the "burst" to it from the edge of Luscombe Wood is the subject of Repton's fourth sketch: his drive runs into the parkland where an ancient oak stands on a projecting knoll and hangs over the

above: The lodges, attributed to Repton, are plain and architecturally subordinate to the house.
opposite: From the top of Back Lawn the sea view was uninterrupted by Dawlish in Repton's day.

castle. A second viewpoint took in the town and church of Dawlish and the sea. From a third, at the west end of the park, the view is surprisingly reduced to sea, hills, and woods. Repton's most remarkable achievement here is the view from the castle itself: from this lower point, the sea provides the distant horizon, the town in the middle distance ("not too much seen, but where the church, & some of the houses, beautifully rising among, & half concealed by the broad mass of the surrounding wood"), and in the foreground "highly dressed Lawn & Pleasure Garden, on which trees & shrubs may be planted, to vary the surface, while they accompany, & partially conceal the distance." Each of these distances was to fade into the next, "making of its several parts one perfect whole: the dressed foreground gradually dying into the Lawn fed by cattle & sheep, from which the Scene will derive all the charms of life & animation." Repton had long sought to bring villages into a composition, and now the town of Dawlish and its parish church became the objects in a view not dissimilar to Ludlow from Ferney, Northampton from Courteenhall, and Corwen from Rûg. However, here there is a degree of sophisticated attention in the gradation of the composition that does not recur in landscape design until Gertrude Jekyll rediscovered it almost one hundred years later.

Luscombe is the most complete work that Repton produced in the second half of the 1790s. In partnership with John Nash, he had learnt that the house need not be a given, a fixed point about which the landscape had to revolve; rather, by adopting the irregularities of Gothic architecture, it could be turned and twisted to become a telling presence in the landscape and, at the same time, to present the best windows direct to the best views. In the accidental valley, without fuss or explanation, he could display his mastery of the English landscape style.

overleaf: Behind the castle, the Back Lawn has steeper, wilder ground than Front Park. The mist rolls down from the moors beyond.

1800–1810

CHANGES IN TASTE

THE TREATY OF AMIENS (March 25, 1802) was a French victory. Food prices fell, rioting abated, and Charles James Fox crowed: "The Triumph of the French government over the English does in fact afford me a degree of pleasure which it is very difficult to disguise."[1]

On May 18, 1803 came war again. France reintroduced slavery, and in 1804, Napoleon declared himself emperor, abandoned the idea of invading Britain, and moved east. There were riots and food shortages in Vienna when he defeated Austria at Ulm (October 20, 1805). The following day, Nelson destroyed the combined French and Spanish fleets at Trafalgar. Napoleon was indifferent. He defeated Austria again at Austerlitz six weeks later and entered Vienna. William Pitt despaired: "Roll up the map of Europe. It will not be wanted these ten years." He died, burnt out and broken, in January the following year, and his death introduced three years of political uncertainty to Britain. In the confusion the slave trade was abolished, but Napoleon was left free to exercise power in mainland Europe. In October 1806, Napoleon crushed Prussia at the Battle of Jena. One month later, his Berlin Decree prohibited all trade with Britain and proclaimed that all the British on the Continent were prisoners of war. The Treaties of Tilsit brought peace between France, Russia, and Prussia in July 1807. Britain was alone.

Although frightening for the British, these sanctions may have been counterproductive. Britain's control of the sea enabled overseas trade in general to grow by nearly 60 percent and re-exports to the colonies by 187 percent. It was also argued that Parliamentary enclosure by increasing the acreage of arable land had greatly increased agricultural output. Now it was decided to embark on a land war, and Major-General Sir Arthur Wellesley, eventually the Duke of Wellington, was dispatched to Portugal in 1807.

In 1805, Wellesley had been recalled to England from Mysore. He had appalled the government by his free-spending on buildings and gardens in India. Uvedale Price speculated to his friend Lord Abercorn that "our governors have been so lavish with our money, and so little disposed to economy, that I fear we must be exhausted even by victories."[2] Nonetheless, he had been a commander in the successful campaign against Tipu Sultan, so he was entrusted with the Peninsular Campaign.

At home, the Picturesque Controversy between Price, Knight, and Repton had destroyed the eighteenth-century consensus, and Repton now began to free himself from his assumption that there might be a single shared understanding of what good taste was. Eventually, in his red book for Stoneleigh Abbey (May 1809), he was to give voice to the heretical opinion that, "the changes in the Taste of a Nation are as little directed by reason, as that of individuals: – there are certain objects which please or displease certain persons at certain times, without any apparent cause; and it would be as absurd to suppose all Men, or all Countries, should think alike on subjects of Taste; – as of Religion, or Politics or any other Subjects."

At one extreme, Repton could search the past history of landscape and architecture for ideas, publishing *An enquiry into the changes of taste in landscape gardening* in 1806, and harvesting the antiquarian knowledge of his son John Adey, the architect. At the other, he revelled in the etchings of India that Thomas Daniell had begun to publish in 1795. These pictures influenced Repton's Moghul designs for the gardens at Sezincote House (1804), and later the Royal Pavilion at Brighton (1806). At the same time, he began to see more value in the Manor house and the cottage, both as architecture in their own right and for the gardens and landscapes that he could design around them.

opposite: The brook at Stanage needed little assistance from art.

STANAGE

— POWYS —

CHAPTER TWELVE

STANAGE

Repton's first visit October 1803,
red book December 1803

THE LANDSCAPE OF Stanage appears to break every rule in the English tradition. There is a house, but it faces east, across the slope, with only one small window to the south and, though now quite grand, Repton had something more modest in mind. There is parkland too—about 600 acres of it—but Repton focused instead on the little brook that runs through it, out of sight of the house, creating a similar web of walks and drives as he had, out of necessity, at Sarsden. It would be right to think of it as exemplary Picturesque, yet, as with Luscombe, there is no burst, nor a pleasure ground such he created at Cobham.

On the face of it, his client Charles Rogers (1752–1820) was far from looking for anything so radical. He was fifty-one years old and the fifth son to a family that had been of some local importance in the fourteenth century but had since declined to a more yeomanly status. Charles, however, had moved to London, made a fortune, and bought himself a large house in Ludlow by the time he was thirty-five. He then began a long, though ultimately fruitless, search for pitcoal near Presteigne, fifteen miles west of Ludlow. In 1799, his fortune much diminished, he still had sufficient funds to complete the purchase of Stanage. Now he was eager, as Repton made clear in his red book, to reestablish his family as respectable gentry in their own country: "We have seen Wealth acquired by industry, succeeding to the hereditary estates of the most ancient families: and we have seen the descendants of those families reduced to want by the vain attempt to vie in extravagance with the successful sons of Commerce; but I have rarely seen, and never before been consulted, in one of those cases where the honest pride of antiquity has been blended with the Prudence and success of commercial Importance: and I congratulate you on the honourable ambition, of reestablishing your family in the neighbourhood, which it has inhabited with respectability for more than six hundred years."

Rogers's search for ancient roots would have perfectly suited the interests of Repton and his antiquarian son John Adey, who was now in business with him, but there is more to the purchase of Stanage than this. Rogers had bought the estate from Richard Payne Knight's cousin Thomas Johnes, who had sold Stanage in order to concentrate his formidable energy on his Picturesque estate at Hafod in Cardiganshire. Repton, in short, was working deep in enemy country, and he acknowledged it: "It will be proper to consider its Situation with regard to the neighbouring scenery, especially as the opposite opinions of two Gentlemen in this neighbourhood, have produced that

previous pages: The house was extended by John Adey Repton and later by the architects J. H. and E. Haycock.
opposite: The steps excluded vehicles from the walk along the east side of the brook.
overleaf: At Stanage, the overriding quality of the landscape is the appearance of a boundless extent.

controversy, in which I have endeavoured to become a moderator"—and he may have challenged himself to out-picturesque the Picturesque at Stanage, as he had at Blaise.

His first step was to dismiss all thoughts of building a modern villa ("new houses like red brick clamps with all the fanciful apertures of Venetian and Pseudo Gothic windows"); nor was it to be a castle. Here Repton couldn't resist a swipe at Richard Payne Knight's house, Downton Castle, "directed by the eye of a painter, rather than that of an Antiquarian." Instead, Repton was to find the antiquity that Rogers sought in the "wild scenery of Stanedge." Repton's design was to be a restoration, an idea he had first introduced at Ferney in 1789, where he sought to restore "the natural form of the ground," but now for the first time, he was to "restore that sort of importance, which formerly belonged to the old Manor house."

Cottages were not necessarily restricted to the tenantry. Charlotte, queen to George III, had hers at Kew in 1771, and shortly after, in 1774, came the ornamental Hameau at Chantilly in France. Repton himself lived in a cottage—in 1810, he was to comment to the diarist Joseph Faringdon that he "... had seen fine places enough, & after all was best contented with his own situation which was by the road side." Manor houses, however, were definitely cottages for the gentry, and Stanage was to be the first of a flurry of such recommendations. At Compton Place, a month later, Repton recommended a design that "pretends to nothing further than the ancient Manor House preserving its Baronial importance by a single Tower." The following year at Longnor, he proposed a manor house where, in the sixteenth century, "the Lord of the soil resided among his tenants, not only to collect the rents, but to share the produce of his Estate with

opposite above: From the house site there was a grand view to the east, down a natural valley.
opposite below: The bridge that carries the north approach is part of a consciously mediaeval setting.
above: Repton proposed extending rather than screening or replacing these old outhouses.

his humble dependents: and where plenteous hospitality was not sacrificed to ostentatious refinements in Luxury." In his red book for Frome House (probably about 1804), Repton acknowledged his debt to his antiquarian son John Adey, first defining the manor house as having "a certain character of architecture, holding a middle station betwixt the baronial castle and the yeoman's habitation," and then recommending John Adey's drawing of Wolterton in Norfolk as a model of the type.

Repton was to work at a number of Norfolk Manor houses in the first decade of the nineteenth century (Mergate Hall in about 1800, Wood Hall in 1806, and Barningham in 1807), however, he had already been thinking about the architecture of Manor houses. Those he admired might be Jacobean, but he described their architecture as "House" or "Elizabethan" Gothic, and emphasised the flexibility of the style in the red book he wrote for Panshanger in 1800: "Gothic Architecture ... of Queen Elizabeth, which has been often abused as incongruous, because in many houses of that date, some bad Grecian ornaments have been introduced" defending their incongruity with the argument that "if the general character be preserved ... good taste which ought to be good sense, will make allowance for those deviations from Antiquity, which are founded on the superior comfort and accommodation of modern times."

above: Repton's pool breaks his rules that "the lively busy intricacy of a running water is not to be sacrificed to the sleepy mirror of a pool," and that "placing water on high ground" was "a material error."

The resulting building would have the stamp of antiquity but could readily be adapted to its setting. So Repton could acknowledge that he was turning on its head the idea of "the modern trim disciples of Brown's defects, that the stables, barns, garden and other appendages, ought to be removed to a distance." A house could be made to look larger by having the stables and kitchen garden attached to it, a point that Repton had already made at Luscombe. It was folly he argued to look for importance in a building "without extent of appendages." In fact, the first house that Repton designed for Stanage was so modest that even he found himself wanting to add a turret to give it a little more weight. He published three sketches of the building in 1802, 1811, and in *Fragments* in 1816, showing his continuing involvement with manor house design.[1] One might say that a line has led straight from Stanage to Edwin Lutyens, via William Morris and the manor houses of the Cotswolds. So, anticipating Lutyens, Repton also advised "straight lines of garden walls and walks, and such management as may form the greatest possible contrast with that rude Character which every where else prevails at Stanedge." Hence, he redeployed that very successful device, which he was also to propose at Endsleigh, of incorporating the walled garden into the ornamental garden, using its south wall as the backdrop for a straight terrace. Walks and drives cut

above: Repton suggested that extra importance could be given by "a single turret," which he added to his watercolour of the house.

into the hill on the north side of the walled garden ran up to the Labyrinth, an "enchanted ground" of a sort, with views out to the north, south, and west.

His cottage lodge, "such as I suppose might have existed at the time of the house, a small orchard behind," was also to reflect the antiquity of Stanage. He had already shown his awareness of the benefit of attaching "gardens and orchards" to cottages in 1801, in his red book for Stratton. Because the house was to be a manor house, neither Gothic nor neoclassical in style, this "small low habitation" made an appropriate lodge. At Courteenhall, he had proposed a rustic lodge with a "rude portico" as a nod to Saxon's neoclassical design. At St John's, on the Isle of Wight, in about 1797, he had attempted another compromise between formal symmetry and rusticity with a matching pair of rustic entrance lodges. This idea he also abandoned. After Blaise he began to turn against the Gothic lodge as well, objecting to the "Pseudo Gothic lodge" at Audley End and dismissing the idea at Brightling in 1806: "I doubt whether any Gothic Entrance can be strictly applicable ... little spruce Gothic ... lodges, too often disgust the Eye of Taste." Yet, the nondescript "natural" lodge, such as he had presented at Babworth, was an unusual way to announce the presence of a great house and a park, unless it could announce to the passerby that here was an enlightened landlord who cared for his people and was not concerned to draw a clear line between the countryside and the parkland within. Nathaniel Kent and Payne Knight had already portrayed the cottage (ideally with a couple of acres attached):

Its roof, with reeds and mosses cover'd o'er,
And honey-suckles climbing round the door,
While mantling vines along its walls are spread,
And clustering ivy decks the chimney's head. [2]

above: The formal terrace on the south wall of Repton's walled garden retains the warmth of the sun into late autumn.

above: The vernacular style of the lodge links the estate to its Tudor past and to the independent yeomen of the Welsh borders.

This was the way to foster an independent class of men, the yeomanry, ready to fight for their country—and no one knew how long the peace of 1802 would last. By 1803, when war recommenced, 380,000 men had joined volunteer militias; by 1805, the number had doubled.

How then was this landscape to be enjoyed? As so often the approach was to run in a straight line across the arable fields north of the house. This road was in place by 1815, but once into the core of the landscape, there would be routes to "rather sweep round, than cross this hill," making for exciting driving, but also to wind past old trees and viewing platforms, those small levelled areas along the drive where a carriage might be stopped for a good view. These, as at Luscombe, do not offer sudden bursts to the house, but views in various directions that recede into the distance. Then there was the reuse of an old holloway—a direct theft of an idea of Uvedale Price's—to run the approach "along the hollow lane in the manner described on the spot with stakes, and if in the execution the banks be preserved, this lane may furnish one of those scenes so often represented by Painters, nor must the sacrilegious spade of modern improvers shave down the steep and rugged banks." This was the approach he had rejected at Cobham.

above: A complex series of earthworks runs north from the walled garden to a levelled platform, where a Labyrinth was planted in the nineteenth century.

Then still more important, there were simple walks along the brook that crosses the park, enlivened with naturalistic weirs, pools, and rocks to give a variety of sound. As for the park itself, this "should be preserved, altho' oeconomy may suggest some means of making it more profitable... there are some parts which might be thrown out and cultivated without injury to the scenery."

"Much of the timber" had been cut down immediately before Rogers acquired the estate, and it seems clear that he was wanting to plant.[3] He may have already sown up to 190 acres, largely with acorns, spruce, and larch. Repton, for his part, was nervous about the predations of the deer and successfully recommended small circular clumps—but he also had in mind the preservation of Stanage's neglected and wild, forestlike character: "... there remain so many birch and thorns and alders mixed with some oaks, that its cloathing is peculiarly picturesque." He was looking beyond economy for those hidden brooks and combes from within which the great house and its parkland would be an intrusive presence. Repton had long wanted to reframe the canvas of his operations as "domain" rather than "park," and here the park, rather than an enclosure of managed grassland outside the forest, was to be rewilded as forest.

above: A grass drive runs off the north approach; a second drive slices across the steep slope opposite.

opposite above: Repton retained such old trees as had survived the improvements carried out immediately before his arrival. opposite below: The fall in the brook, designed by Repton, has survived. above: A simple clapper bridge across a low fall of water makes a striking contrast with the reputed elegance of Regency taste.

BRIGHTLING
— EAST SUSSEX —

— *CHAPTER THIRTEEN* —

BRIGHTLING

Repton's first visit March 1806,
red book June 1806

Repton's red books sometimes read as unsubtle exercises in flattery, as in 1800 when he wrote, "I cannot better preface my opinion for the improvement of Hatchlands, than by paying a deserved compliment to the good taste of its possessor." But such ingratiating comments were always an instrument of diplomacy. Where his personal opinions do break through, as they did in his red book for Hewell Grange (January 25, 1812), and as they tended to against his Picturesque critics, Price and Knight, they are all the more shocking as a result: "They attack the Art itself & under a battery directed against the late Mr. Brown, they attribute to him & even to myself much of the bad Taste of his followers."

Repton was in an odd position at Brightling. Capability Brown had drawn up a plan for the landscape "to increase the beauty and the value of the land at the same time, by manuring, by draining, and above all by sheltering it from the cold winds of the North and East." In 1788, Capability Brown's oldest son, Lance, had married Frances Fuller, the sister of Repton's client John Fuller III (1757–1834). It might have been a mistake for Repton to criticise Brown's work on that account.

On the other hand, Brown had been dead for more than twenty years, and his style was under attack by Price and Knight. Repton had to be critical to get the commission: "... some parts of Mr. Brown's plan having been misunderstood, whilst others have (fortunately) not been adopted."

Fuller had taken possession of Brightling thirty years earlier, but only now was he ready to spend (he owned plantations in the West Indies, and was probably repatriating his wealth because the slave trade was about to be abolished). Repton's first call on Fuller's pocket was to propose a new house in the middle of the park, which he could give his son John Adey to design. Rather than blame the family, or history, Repton (quite unjustly) targeted Brown for the failings of the old house: it was too high up, and as a result was "exposed to the prevalent winds," and though it had a "vast command of views," these views were all prospects, a difficulty he had touched on at Blaise and defined at Tendring in 1791: "All inequalities of surface are lost by looking over them."

Accepting that the house might not be rebuilt, he made modest proposals to settle it better into its landscape with some architectural adaptations and a flower garden on its northeast side. He also proposed works, which were carried out, to the Avenue, a belt that Brown had planted with beech and Scots pine, running out from the house along the

previous pages: Away to the north, the woods and fields of Sussex flow over rolling hills.
opposite: Sir Robert Smirke's summerhouse beside the terrace survives,
put up three years before Repton arrived.

ridge close to the north boundary of the parkland. There was already an entrance here, but Repton was to offer a new line for the approach with ha-has on both sides. Double ha-has were an expensive solution, particularly as these had elaborate swellings in their line. Such swellings occur nowhere else in Repton's work, and they are unnecessary. In fact, it seems likely that he intended the ha-ha to follow the simple line of Brown's wood with "projecting masses and groups of trees" planted in front, but his instructions, like Brown's, were misunderstood and the ha-ha was carried round the outside of his projecting masses.

Repton adopted a route that slipped from one side of the ridge to the other so as to show off the views on both sides, beginning at Avenue Lodge ("The Terrace commands an extensive Inland View towards the North"), then offering views across the parkland towards Beachy Head ("The Terrace commands an extensive view of Land and Sea towards the south"), before ending, as the approach made its final turn and dipped towards the house, with a splendid delayed burst, showing two fronts of the house through a tunnel of cherry laurel. For the rest, there is a good deal that is Brownian in the park. Besides the walled deer park, there was to be a paddock in the far southwest corner of the domain—a good distance from any new tree plantations, though in view from the house. Repton sought to retain a Hare Park, though the evidence suggests that it was built in the southeast corner rather than, as he proposed, in the northwest. His walled garden was built, with a Brownian pleasure ground around it, and there was a good deal of grass. The park had been broken into

previous pages: Repton wanted to build his new house on the knoll where the Temple now stands.
above: The house at Brightling, known to Repton as "Rose Hill," is comfortable but unremarkable, and is faced in red brick, which Repton abhorred.

three in Brown's style, and there were occasional buildings, some now lost, but including at least one lodge, a rustic seat with "posts ... covered with creepers in the foreground" to create "three distinct landscapes," and a fold yard, all of which were in place by the end of the year, as were the alterations to the Avenue.

On a subsequent visit in 1807, Repton had accepted that the house would not be moved: "The House is too good to be puled down – & too bad to warrant much expence in alterations & the Inconveniences of the present Situation have been already stated," but he and Fuller set out a "new line of approach from Lewis." This was to enter at the southwest corner of the deer park, then find its way "mark'd across the dell" from episode to episode, across a bridge and onto the floor of an "accidental" valley with, at its head, a burst to the Temple, after which a causeway beside a sunk hedge carried it up through the pleasure ground to the house.

Earthworks in the dell (a wooded stream-fed dingle). suggest that this had an intricate layout, but Repton also selected it for "the boundary of the Deer Park." His stone wall would thus have been hidden—the public road between the house and the Hermit's Tower was sunk to conceal the boundaries of the parkland in the same way.

The wall excluded the site of Furchase Wood from the park, and Repton proposed its planting, clearly intending to thin it into the parkland "The best mode would be to plant the whole, and afterwards cut it into shapes; it might then be considered a wood interspersed with lawns," for his ideas were moving on from Courteenhall, Luscombe, and Mulgrave Castle where he had recommended "a sweeping continuity of wood, connecting the hills and vallies." Now he had in mind "the outline of Stoken Church hill ... a forest partially cleared of wood by time and accident." It had become essentially an extensive wood pasture, though tailored to address

above: As he had at Courteenhall, Repton planted small clumps with mixed species to break the line of the surrounding tree belt.

Had the house been moved closer, as Repton wished, the dingle at Brightling, mysterious and alluring, would have been his "enchanted ground."

In his early career, Repton would never have suggested something so simple as this bridge on the Lewes approach.

small changes in the topography. Repton appears to have planned something similar for Stanage, and pastures like these survive at Moccas, Ashridge, and Cobham. Repton generally held the square fields of Parliamentary Inclosure with their thorn hedges to be at odds with his love of natural landscape. At Normanton in 1797, he compared unenclosed wasteland and old hedgerows with "newly inclosed" land, acknowledging the increased yields of the latter, for which his friend Nathaniel Kent had argued, but also objecting because: "There is a wildness of character in an open heath country, which tho' not so profitable is certainly more picturesque than the straight lines of hedges and square fields, which like a draft board, chequer the whole face of a country after it is first inclosed." For Repton—besides being more beautiful—the old hedgerows could provide firewood and nurseries for game; they offered shelter and fertility, and crooked timber that would answer the needs of the Navy. At Stratton, four years later, he amplified the argument, recommending "those lesser masses of brushwood and hedgerow timber, which in more enclosed countries form a sort of continuity of wood" and proposing to replace the arable fields around the village with "a sub-division by live hedges and fruit trees, making small enclosures of two acres to each cottage for their gardens and orchards" to "give comfort to the inhabitants" and make the landscape more beautiful. Enclosure would impoverish the poorer villagers, stripping their rights to common grazing and common fields. Like many, Repton thought this would be followed by the loss of the traditional independence of the yeomen, of community, of identity, and of the look of the country where ancient hedgerows, up to 100 feet wide, would be uprooted. These people might lose their land and homes, though these had been owned by the same family for centuries, leading to rural depopulation and political division.

On the other hand, enclosure allowed for land to be gathered into large units under one ownership and its efficiencies made enclosed land attractive in a time of war, with high prices for food and wool. At Brightling, therefore, following the General Enclosure Act of 1801 (whereby any parish was allowed to enclose its land if three-quarters of the landowners agreed), Repton returned to enclosure with more sympathy, describing "a View of large tracts of open country recently inclosed; where the lines of hedges are often traced on the map by the commissioners at

opposite: The Temple, designed by Smirke in about 1811, is the pivot about which the landscape revolves.
above left: Cobley Lodge has land around it and an air of independence denied to buildings in an obviously classical or Gothic style.
above right: The square holes in the park wall are "mewses." Hares, lured through them by food, were trapped in the park when the mews were closed, and made the prey of dog races.

opposite above: The Hermit's Tower was used by the Keepers on the estate.
opposite below: Smirke's Observatory is a beautiful building, too elaborate perhaps for its outlying position in the landscape.
above: Sugar was formed into cones before being transported. There is a story that Brightling's sugar-cone was put up overnight to win a bet as to whether any church steeple were visible from the windows of the house.

right angles; and in the most perfect squares ... but generally there are such inequalities of surface that ... every square field takes a different shape to the eye, although exactly similar on the map."

At the same time, the sunk road and park wall (reputed to have cost a cool £10,000) show that Repton was still applying himself to the wider countryside, and it was here that Fuller employed the neoclassical architect Sir Robert Smirke from 1810–1812 to design the Observatory (showing the breadth of Fuller's scientific interests); the Needle, the Hermit's Tower, and the Pyramid (his cultural sensibility); and notoriously, the Sugar Loaf (a proud acknowledgment of the source of his wealth). There were to be no hard boundaries. Furthermore, with these buildings, as with the Temple inside the park, Repton had learned to celebrate "the different Characters of buildings in the same place," which he had reluctantly accepted at Blaise as "in Music, the modulation from the major to the minor key"—a metaphor that suggests that he could let go of architectural unity so long as the big house remained dominant.

Perhaps though, for all Fuller's money, Brightling was not quite the success that Repton would have liked it to be: his proposals largely relate to the edges of the parkland; the dingle was not developed to match Endsleigh or Stanage; and the central core, around the Temple, remained undeveloped. Perhaps he could not finally relinquish his ambition for a new house.

right: The Pyramid Mausoleum was built in the graveyard of Brightling Church in 1811; Fuller was buried there in 1834.

1811–1818
THE LATE WORKS

In 1811, George III went irrevocably mad and his son was made Prince Regent. Facing unemployment and famine, the Luddites smashed the machines of the industrial revolution, fearing that machines would replace them. From 1812, what had been European became a world war. The United States declared war against Britain and seized the opportunity to raid Canada. Napoleon invaded Russia because it refused to "act as my second in my duel with England." His retreat marked the turning point: in October came a decisive victory for the allied armies at the Battle of Leipzig.

In January 1811, Repton had a carriage accident and injured his spine. He never fully recovered, and by 1814, he was confined to a wheelchair. He was thought to be close to death in 1815, yet between 1811 and 1814, he produced his finest and most joyful works. He reintroduced parterres and terraces, and designed strings of discrete garden areas, each with a distinct character, often harking back to a Tudor or late-mediaeval past.

This was not a response to military victory. Napoleon was not exiled until 1814, and was only finally defeated by Wellington and Blücher at Waterloo in 1815.

opposite: The bounding walls on the west side of Pentillie Castle long pre-date the Reptonian American Garden within.

PENTILLIE CASTLE
— CORNWALL —

CHAPTER FOURTEEN

PENTILLIE CASTLE

Red book February 1810

THERE IS EVERY reason to suppose that Repton was delighted with his commission at Pentillie Castle. This was an ancient house on a magnificent promontory high up above the River Tamar in Cornwall. His clients, the Corytons, were a long-established county family, and the estate was substantial. Repton had been recommended by a mutual friend, Reginald Pole Carew, his old client at Antony.

Here was another opportunity for an expansive Brownian park. Repton himself admitted as much: its "interest," "beauty," and "apparent magnitude ... would justify a rivalship with Belvoir, or even Windsor Castles." In the same breath, however, he acknowledged that he saw its "commanding Scite" as presenting "difficulties in its treatment." Pentillie was an architectural muddle: only its two turrets and the tower gave "to the whole mass a degree of Gothic importance." It looked out to the east across the Tamar, while the land most suited for emparkment was largely out of sight behind the house, to the west. Also, because it stood on the end of a promontory, it was necessary to have the land approach and entrance on the same side as the pleasure ground.

Besides the red book, two other accounts shed light on Repton's design. First, Pole Carew sent him a detailed brief in 1809. Then, following the carriage accident that disabled him in 1811, Repton appears to have revised his red book, and, in 1813, to have handed the commission and his revised design over to the nurseryman and designer Lewis Kennedy, who wrote his own "Sketches & Notices" for the place in October 1813. The handover is implicit in the comment with which he concluded the red book: "When I reflect on the distance of Pentillie Castle from my usual scenes of action, I feel the impossibility of explaining all the improvement of which its picturesque features are capable."

Carew made several points in his letter of 1809. First, he recommended the view up to the castle from the river: "I as one of the Public ... think it cannot be improved." Next, while the massing of the building might be good, Carew thought the building itself unacceptable: "It can hardly be said to belong to any style of architecture, and the Details both inside & outside are extremely bad." He then recommended removing the hedges on the west side of the house, wanting "only the Hand of the Master to connect & form them into one interesting whole." He advised replanting the steep falling ground on the eastern river side with low brushwood, to replace an old orchard that had already been largely replaced with "one or two circular Petty-Pan Plantations."

previous pages: Today the castle's river front is more or less symmetrical.
opposite: Repton planted the steep slope below the ancient castle with oak.
overleaf: Pentillie still commands the long reaches of the Tamar river.

Carew's suggestions were in step with Repton's work of the 1790s, when he was at Antony, and Repton, whose practice had evolved considerably in the intervening seventeen years, showed some diplomacy in the way he handled Carew's advice. Carew was anxious to protect the impression that strangers ("the publick") would have. Repton had deployed the eye of the stranger as an arbiter in his earliest red books, as we have seen at Cobham, but in 1791 at Nacton, he had confessed that "I am not only to delight the eye of strangers who may visit Nacton, but I am to make the home of a friend convenient and comfortable." So at Pentillie, although the strangers were confined to the river, he could still argue that "the comfort of those by whom it is to be inhabited ought not to be sacrificed to the publick," and while he "might have suggested plans more beautiful and more worthy of the natural Situation of the place" (such as would appeal to strangers in search of beauty), he regarded it as his duty "to confine myself within such limits of prudence, oeconomy and propriety, as may render my advice practicable."

On the opening pages of the red book, Repton agreed to remove hedges as Carew had proposed: "... the adjoining hills now intersected with hedges, may with propriety be laid down to grass, to form an

above: The extensive tidal marshes of the Tamar were not drained until 1902.

adequate extent of Park or lawn annexed to the Castle." However, further on, he qualified his agreement with a sneer: "The effect of creating extent by taking away hedges and converting the cornfields ... into one large Sheep walk; is such as may be done any where, and will be deemed a great improvement by all who consider the removal of fences as the basis of modern Gardening." This later page seems to have survived from an early, more radical draft, critical of "modern gardening" (by which he meant Brownian design), wherein he sought to turn his client's attention to the "enchanted ground" of Pentillie—"the scenery near the Root house, and in the Woods, and about the Quarries." This was one of the projects that Kennedy was to undertake, creating a sequence of gardens off both fronts of the castle. In fact, Repton hardly saw the need for any parkland, making the point in the first part of the red book that "It is not expected that a Castle should be situated in the middle of a park." He reined in the ambition of his clients for a large park, and this time offered the further justification that "the well-chosen scite of a Castle supposes it to be difficult of approach, and a sloping lawn presents a most inviting parade for an army to march to its attack." The landscape was to conform to the castle style of the building, hence he repeated the approval

above: In order to maintain the semblance of antiquity, Repton retained the guard house at the entrance to the pleasure ground.

he had given at Stanage for wildness. He knew it was not parkland but did not have a name for it, advising the "same covered with brushwood ... as one thicket of thorns, holleys and nut bushes" and noticing "some walls and steps, that contribute to extend the same Character quite from the Castle to the banks of the river, and will furnish an immediate and easy connection between these two objects," and "the walls, the deep hollow ways and defiles near such a building, are in perfect harmony with its Character."

Again, he modified this idea in his plan "HINTS FOR DISPOSING THE GROUNDS NEAR THE CASTLE (Repton's capitalization)," which seems to have been another later addition, where he recommended "This Steep Bank to remain as it is with Orchard Trees Filberts & flowering Shrubs intermixed," and above it, on the more level ground immediately east of the castle, a grazed lawn or, ideally (as actually happened) "a Rosary, or flower garden ... enriched by large masses of Shrubs intermixed with fruit as Gooseberries Raspberries &c."

The presence of a castle in the Middle Ages would have encouraged tradesmen to set up in the vicinity, taking advantage of the protection it offered, and

above: The west front at Pentillie, less grand than the river front, is also the entrance.
opposite: Repton's revised plan shows the west lawn and the east front much as they are today.

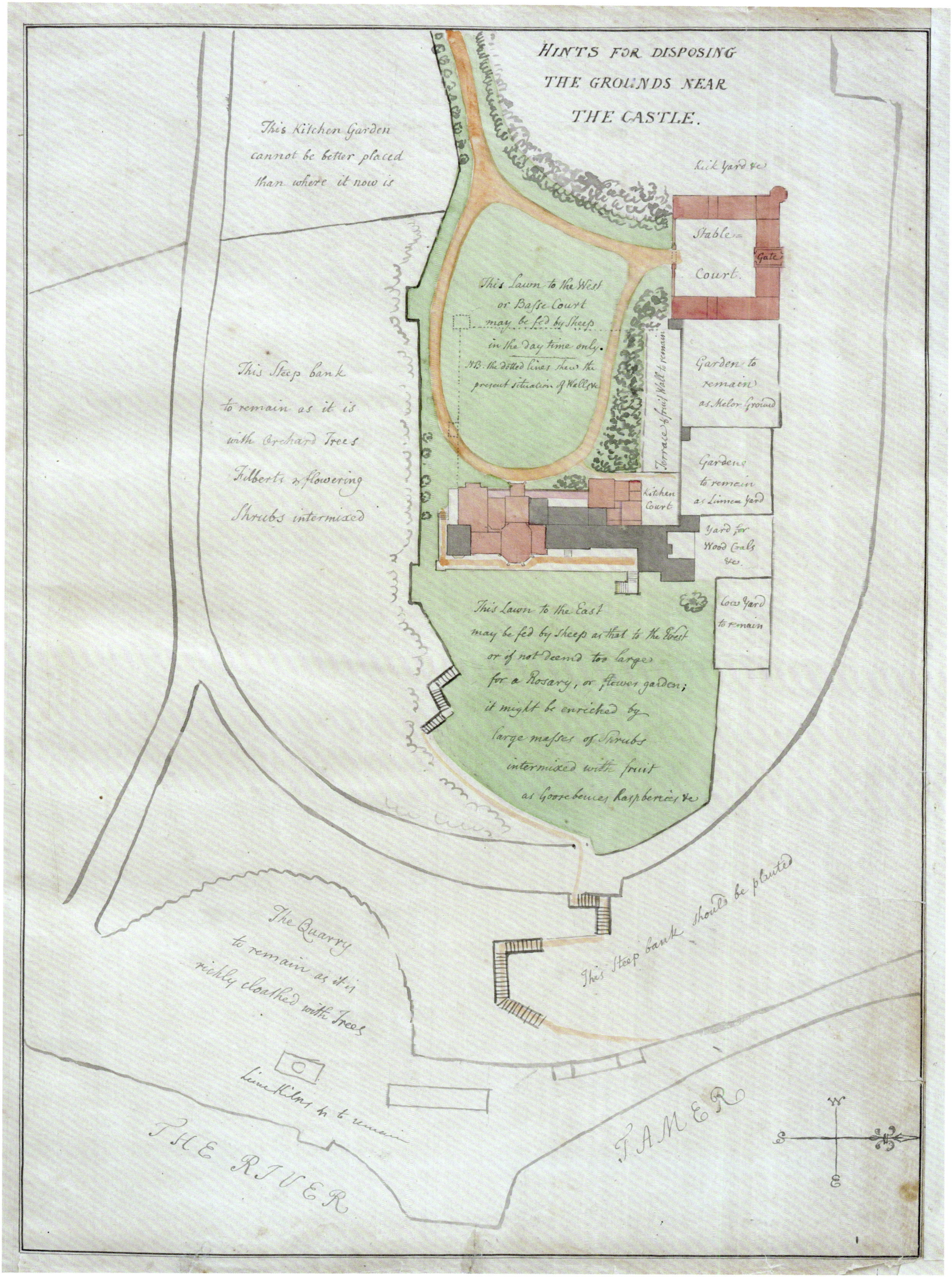
HINTS FOR DISPOSING
THE GROUNDS NEAR
THE CASTLE.
This Kitchen Garden
cannot be better placed
than where it now is
Rick Yard &c
Stable=
Court.
Gate
This Lawn to the West
or Basse Court
may be fed by Sheep
in the day time only.
N.B. the dotted lines shew the
present situation of Walls &c
Terrace & fruit Wall to remain
Garden to
remain
as Melon Ground
Gardens
to remain
as Linnen Yard
Kitchen
Court
Yard for
Wood Coals
&c.
Cows Yard
to remain
This Steep bank
to remain as it is
with Orchard Trees
Filberts & flowering
Shrubs intermixed
This Lawn to the East
may be fed by Sheep as that to the West
or if not deem'd too large
for a Rosary, or flower garden;
it might be enriched by
large masses of Shrubs
intermixed with fruit
as Gooseberries Raspberries &c
This Steep bank should be planted
The Quarry
to remain as it is
richly cloathed with Trees
Lime Kilns &c to remain
THE RIVER
TAMER
W
S
E

The lodge (above) has been extended and reroofed,
but it remains a reworking of Repton's idea in the vernacular tradition (below).

top: Repton proposed leaving Pentillie Castle as an unashamed muddle,
but recommended whitewash to unify its architecture.
above: To the east, the horizon is bounded by the barren lands of Dartmoor.

This Monument is erected
In Memory
of

Repton therefore sought successfully to retain the lime kilns on the river side, as Carew had recommended: "I must direct the attention to the buildings at the foot of the hill, as tending to mark the height and importance of the Castle." In the same vein, Repton gave careful instructions for a lodge in the cottage vernacular: "A building in the Style of the Castle, at such a distance from it, would not be so appropriate as a mere Cottage of coarse granite & and covered with reed straw thatch—for which a hint is here subjoined; but much of its picturesque effect will depend on the honey-suckles and creepers with which it should be decorated."

The last parkland building to concern him was the Mausoleum, "The small building on the apex of the hill, called Mount Arrarat." His concern, akin to Carew's, was that "a stranger might suppose it to be a Summer house of some wealthy citizen, who had a villa in the valley beyond." In order to settle it into the castle landscape, he proposed that "By the addition of a small tower at the corner, and the appearance of a wall or terrace, it might easily be converted into an apparent outwork, and become a part of the place." However, this building had a history. It was a two-storey summerhouse, in which Sir James Tillie had had himself interred when he died in 1713. He was strapped to a seat, with servants to bring him food every day in case he revived. His bones were not removed to the parish church for a century. Repton's solution, which seems to have been carried out, was to bury Tillie's resting place under a terrace, and add only the external walls of a third floor to give the effect of the "isolated Tower of a ruined Castle."

At Warley in 1795, Repton had ruled that "a house in the Country ought not to be very lofty." So, at Honing, he emphasised the horizontals of the house by inserting a cornice. If he could bury the ground floor of this summer house, Repton could equally well "bury the basement story" of the castle, as he had at Welbeck in 1790, at Bulstrode and at Catchfrench in 1793, and at Rûg in 1795. This would bring the principal rooms down to ground level and give the house a more modest place in the landscape. Consequently, he proposed to build a terrace along the east front "the whole length of the building," which would also make space for a large formal flower garden (hitherto unsuitably placed at the dusty entrance on the west front).

Lewis Kennedy's "Sketches & Notices" must be read as his interpretation of Repton's instruction. The notes describe a series of gardens, all of which had been introduced elsewhere by Repton: the Thornery, which he had experimented with at Woburn; the children's garden, which he was to produce again at Endsleigh; the quarry garden, such as he had developed at Catchfrench and Port Eliot; and the flower garden. This is the same proliferation of garden types as Repton had experimented with at Sarsden, and would propose at Ashridge, Cassiobury, Endsleigh, Montreal, and at Wanstead, where Kennedy again finished the work.

Last but not least came the approaches through Bittleford Wood with their dramatic views over the river and up to the castle. Repton congratulated himself that "The Approach thro' what may be called the natural part of these grounds, will be highly approved by those, who think every thing belonging to a place should be natural and picturesque." But these approaches would have been expensive to construct, and they gave him a greater challenge than he had faced at Sarsden and Blaise in bringing the approach up the steep hill to the castle. They did, however, have the benefit of presenting the castle in silhouette only. Repton agreed with Carew that it had been built in several phases and with a "total difference in the styles of Architecture." In the revised design that his client accepted, he recommended leaving much of it "because it is the general effect, and not the detail, of a Castle which it is necessary to preserve" and "a want of uniformity always increases the apparent extent, and importance of the Castle Character as much as the irregularity of its outline."

opposite: Sir James Tillie, though in stone, is still seated, still awaiting resurrection.
overleaf: Repton created what looked like a ruined outpost of the ancient castle by burying Sir James' summerhouse under a terrace.

ENDSLEIGH COTTAGE

— *DEVON* —

CHAPTER FIFTEEN

ENDSLEIGH COTTAGE

Repton's first visit August 8, 1814,
red book September and October of the same year

ENDSLEIGH COTTAGE, in name and in the quality of his work here, does not just bring Humphry Repton's career to an end, but to a perfect end; and Repton's place in the Pantheon of English design might be assured by this landscape alone, for all those difficulties that faced him in his career somehow find their place of rest in this remote part of North Devon.

Repton had been working for the 6th Duke of Bedford at Woburn Abbey since at least 1804. The duke then decided to build a fishing lodge nearly two hundred miles away on the upper reaches of the River Tamar. In 1809, Repton and his two sons visited and drew up three sketches of proposals. This was the year in which Repton was developing his plans for a manor house at Stanage. Two of these sketches, the most radical, went further still, in planning to extend "an irregular Farm-house, little better than a Cottage, backed by a hill and beautiful groupe of Trees," with a string of chalets in the same cottage vernacular. Though rebuilt, at first as a cow barn and now Pond Cottage, this is a thing to wonder at—for this Endsleigh was to have been a village refashioned, not like Blaise Hamlet for the labourers on the estate, but as a ducal residence. Cottages were now to take centre stage.

Repton avowed that the "good taste of the Noble Proprietor" had chosen the site, describing it first as a "groupe of Rural buildings" (the Reptons' first idea), and immediately afterwards as a new design "such as will do credit to the Name of Wyatt" (referring to the Cottage itself, eventually built on higher ground, to the east).

Construction of the Cottage and its setting started soon after 1809, and it has been claimed that both were the work of the architect Jeffry Wyatt because Repton dated his red book "Augt 8 1814."[1] However, the third of the Reptons' 1809 sketches was for a new house on the site eventually adopted for the Cottage, and, after his accident in 1811, Repton was retained by the duke. It is likely, therefore, that Repton and Wyatt collaborated from around 1810 (as they were to do at Ashridge in 1813), and that Wyatt took charge of the execution of the works until August 1814, when Repton was encouraged by the 6th Duke of Bedford to make a second visit. The style of Endsleigh is so entirely late-Repton in its disregard of the rules of design that no other conclusion would be readily credible.

The Castle at Luscombe broke the landscape into two parts; Endsleigh Cottage breaks it into four. A long approach through dense woodland provides the

previous pages: Repton levelled the lawn using the material cut from the hill to make his terraces.
opposite: Only from the burst, a rocky point above the Tamar, can one see how well Repton accommodated the cottage to its setting.

introduction. It has no view of the river and no burst to the house. Instead the back wall of the pleasure ground appears on the left and, moments later, a clutter of drain pipes and a dismal back yard at the north front greets the visitor. Approaches, once a centrepiece of Repton's design, have here been sacrificed to the views from the Cottage. Then, stepping through the Cottage comes the staggering view east, to the terraces of the flower garden, the lawn, and the long reach of the river; and turning to the right, a third landscape, with the parterre of the Children's Garden in the foreground and the hills on the Cornish side of the Tamar behind. To find the greatest of all these scenes, the "Dell West," however, it is necessary to walk out two-hundred yards from the Cottage. This is not a landscape that flows; on the contrary, it has different components, but the position of the Cottage makes the whole harmonious by separating one from the next with no appearance of effort.

Suddenly, the architecture needs to be reassessed. It is cottage vernacular, taking full advantage of the freedoms of manor house design that Repton had explored at Stanage. A cottage can sprawl; a cottage eludes any attempt to subject it to the rule of "taste." On the contrary, a cottage offers complete freedom in the massing of its parts (in form, Endsleigh Cottage consists of two buildings linked by a corridor, akin to the greenhouse he had proposed at Babworth and his first idea for the "irregular Farm-house" at Endsleigh). It is characteristic of Repton that even before the Cottage had been completed, Repton himself was questioning this new style, writing in his red book for Sheringham (July 1812), "The Character [of Sheringham Bower] is not to be a modern pseudo Castle, or Abbey: neither is it to be one of those Large Mansions, which Modern Fashion has been pleased to call Cottages."

It is not surprising, given his own lack of self-awareness, that Repton should so underestimate his achievement. Luscombe, Moggerhanger, and Stanage all have elements that one can see in Capability Brown's work, albeit further developed, but Endsleigh is so far from Brown that it will be better to think of Repton as starting afresh with the art of landscape. More radical than Moggerhanger with its successful integration of public and private space, Endsleigh has Sarsden's precise control of views. More radical than Stanage—where Repton conspicuously turned his back on the idea of Brownian parkland, but in the end his Manor House was always baronial—at Endsleigh, there is no pretension, and not even the suggestion of a park. The wedge-shaped lawn that runs out from the Cottage at an angle, may have a precedent in the Quarter-deck Lawn at Mulgrave Castle, but at Endsleigh, it is presented with infinitely more strength. It would never have worked on a drawing—small wonder that Repton had to make a second visit to rescue Wyatt. Then, hanging above that lawn, there are garden terraces, as though Repton were finally making amends for those that he planted over on the equally steep slopes of Ferney and Blaise, and running along the lowest of the terraces, herbaceous borders, "borders of curious flowers," such as Repton had introduced at Valleyfield in 1801. These would not be seen again at this scale until the time of William Robinson in the 1870s.

Hidden around the corner of the Cottage was the parterre in the Children's Garden. As late as 1807, Repton had showed at Woodhall "... the interior of the flower Garden" as a set of amorphous island beds, but suddenly in 1811, at Uppark, he was writing about formal gardens in terms that might have come from an arts and crafts handbook: "the beds might be edged with a very deep box clipped to two feet high," and in all the late works, Ashridge, Wanstead, and Endsleigh, the parterre recurs. One can criticize Repton's work at Pentillie and Ashridge for creating gardens in a series of little private cells, cut off from each other ("The gardens, or pleasure-grounds, near a house, may be considered as so many different apartments belonging to its state, its comfort, and its pleasure," as he put it in his red book for Woburn Abbey in 1805). Being so concerned to keep each experience distinct, he put each into a separate room; but this is not a criticism to be levelled at Endsleigh, where the various designs respond directly to the topography.

previous pages: Repton had always felt a tension between the artificial and the natural; yet in his "Dell West" he fused the two.
opposite: Repton had Jeffry Wyatt's retaining wall and bastion removed.

No X

No X

Repton's picture seems remarkably to herald the Arts and Crafts Movement. It is equally remarkable that the scene survives and is immediately recognisable today.

Among all the great things that Repton achieved at Endsleigh however, the "Dell West" is surely his most impressive and most completely achieved enchanted ground. This has everything—the brook, exposed rocks, winding paths, surprise views, all orchestrated with the sound of countless waterfalls and birdsong. There is a hint of the design he sought here in the red book he wrote for Valleyfield in 1801: there were to be "rare plants of every description ... beds of bog earth ... for the American tribe ... aquatick plants ... on the surface, or near the sides of the water, and the numerous class of rock plants ... [with] rugged stones provided for their reception, without the affectation of their appearing to be the natural production of the soil. But above all there should be poles & hoops for those beautiful creeping plants which form themselves into natural festoons when supported by Art & attention. Yet with all these circumstances of beauty, the flower garden should not be visible from the roads or general walks about the place ... & its decorations should be as much those of Art as of Nature."

The passage bears quoting because Repton's written advice for the dingle was so concentrated on the water—indeed, the dell does not need a gardener today so much as a conductor, both to lead the water to its channels and to conduct the music that it makes as it falls down in eight or more rivulets, each with its own range of sounds. Repton had long sought this, as was apparent in his red book for Stanage: "... the different sounds produced by such water-falls, will add to the variety and sollemnity of the scene; from the gentle trickling of a small rill, to the dashing roar of a more copious stream, varying with the quantity of water and the height of the falls."

There are bursts too, but they are not from the drives, or indeed the approach. The true bursts to the house are first from a point of rock accessible only by a steep and perilous footpath, just south of Wyatt's Swiss Cottage, from which there is one view down a long reach of the river and another back to the house. The second view is from the far side of the Tamar, where Repton proposed building a "Corn Mill." He had played with such bursts before—Jupiter Hill at Antony is the pivot

right: The parterre garden is replanted at least twice a year, but is never without flower, and has not changed in form since Repton designed it.

around which the landscape revolves, and it too is only accessible along a narrow path.

This mill was to be prominently in view, its Picturesque effect enhanced by "the occasional traffic and busy motion of persons crossing the Tamar" to reach it. It was to be powered by a cascade—an idea that he had put forward at Burton Park in 1798, and again at Aston in 1802, but in both those sites, as at Babworth, the mills themselves were to be concealed. Not until Stoneleigh in 1809 did Repton acknowledge that "the old abbey Mill and Bridge" formed "an interesting part of the peculiar Character of the place."

At Wallwood, Repton had recognized "the cheerful moving scene of a public road," only cautioning "not too near to be offensive." At Endsleigh, he aimed to put all the machinery of landscape on view—not just the corn mill, but also the kitchen garden and the hard landscaping of the terraces—and to animate the scene further, in the small fields below the Cottage, a milking parlour. The result was a new kind of landscape, which rejected the expectations of a country house.

The contradictions embedded within his late red books suggest an increasing impatience with the constraints of English design. The modest treatment of the building, the acceptance of haphazard organic architectural growth, the reluctance to make any kind of parkland—these show an ever-increasing reliance, perhaps promoted by his son John Adey, on truth: truth to the appropriate in a setting, truth to the vernacular, truth to a romantic sensibility that would be stimulated by the series of distinct gardens, and a final reliance on truth to the place.

Visitors looking at Endsleigh today may not readily see how remarkable it is, partly because it is not so different from a garden of the Arts and Crafts Movement, such as might have been made at any time in England over the past one hundred years. But Endsleigh was made two generations earlier: Repton popularized an ideal in his descriptions of Endsleigh and other of his late gardens in *Fragments*, and in so doing, he established a new style, enthusiastically adopted throughout the Anglophone world.

above: A water-filled runnel on the balustrade in Repton's Children's Garden allowed children to sail their toy boats.

above: Repton did not mention the Swiss Cottage in his red book, but must have planned a building here, where one could take refreshments and enjoy the view.

*above: The entrance lodge is a "natural" design,
with a substantial back garden. It has none of Blaise Hamlet's French fancies.
opposite: The background drone of a continuous fall of water—the fluting laughter of a shallow glide over loose stones, the tenor chuckle of single rocks when water tumbles over them, and the bass note of a fall onto a hollow—all these sounds combine and recombine at every step along every path.*

opposite: Pond Cottage stands beside the stream at the foot of Repton's Dairy Dell.
above: Repton's dairy was central to the design and very much a working building.
overleaf: Seen from the north side of the stables, the jumble of roofs
on the north front give no foretaste of the landscape.

ENVOI

In those interesting years that fell between the fall of the Bastille and the Battle of Waterloo, the old order in Europe was shattered and it has never since been united. Many histories have been written about the period, and Repton's career provides the stuff out of which more might be woven. Yet through these twenty-four years of tumult, Repton steadily worked towards a set of original solutions to the way we live, solutions that are so conspicuously present in our world today that we take them for granted.

William Kent had trained as a painter, Capability Brown as a gardener, but Repton's undisciplined eye saw things that escaped those whose upbringing was more deeply embedded in the English tradition. His method was scientific: he would adopt existing ideas and, where he found them wanting, he would try others, and would continue trying until he had come to a satisfactory conclusion. Because he was an original and an artist, rather than a social construct, we must therefore excise the following six true and tempting, but ultimately belittling, social propositions from any study of his art:

I

The rapid increase in corn prices during the French wars encouraged landowners to promote enclosure and the engrossment of their estates, and to farm in hand. Consequently, they acquired large dependent bodies of workers available to build and plant the landscape and had no need of contractors (such as Capability Brown had been). As a result, Repton did not have full control of the execution of his design.

II

The rise of an independent professional middle class, of which Repton was a member, brought benefits, enabling him to charge for his services without loss of self-respect. With the rapid growth of industry and of the cities that supported it, it was sensible for this class to live in villas with easy access to town. Villas inevitably had relatively modest estates and so limited the scale on which Repton could work.

III

Rising with this villa-dwelling, city-based middle class came a new breed of politicians, largely drawn from the growing financial sector—people for whom landscaping was not about aggrandisement, or perfecting nature, but about making places in which they could comfortably live.

IV

At a time when the French revolutionaries had declared that there was no God, a love of Olde England and church-going were proofs of patriotism. Many of Repton's clients were evangelical, driven by a desire, which Repton shared, to live better lives themselves, and to

opposite: With Repton's Blaise, as with Capability Brown's work, the English style frustrates the attempt to distinguish art from nature.

improve the lot of mankind—hence, their inclination to build cottages for their dependents.

V

While the war and Pitt's income tax reduced the purse that Repton's clients had available to expend on their estates, it remained a patriotic duty to plant oak, for Britain's position depended on shipbuilding and supremacy at sea.

VI

Repton was a child of Romanticism, which drove him away from the precepts of his master, Capability Brown, and away therefore from any visionary transformation of the fabric of England, and towards satisfying the material needs, and the comfort and convenience of each client.

Each of these narratives casts Repton as a product of his time and his society, and all may be true. Yet each wonderfully avoids the celebration of what he actually did and what was original in his work. It may be a trite observation, but time and society do not act to make the artist; the artist, acting on time and society, makes the art. Let us clear these propositions off the table, therefore, and start afresh.

Another measure of the artist's endeavour, equally beguiling and equally delusive, is to evaluate its influence and legacy.

VII

Many people nowadays will say that Repton's red books, rather than his landscapes, are his legacy—these beautifully painted and bound accounts of landscapes as he found them and as he proposed to transform them. In our own day, the great Swiss scholar André Rogger has forcefully argued that they should be regarded as works of art in their own right.

VIII

More influential than his red books are Repton's published accounts of landscape theory. These found their way across Europe and the United States, and directly influenced the great designers of later generations, including Prince Pückler-Muskau in Germany and Frederick Law Olmsted in the United States.

Let us clear these books off the table as well: the red books are far from a final account of Repton's ability as a designer. Their formulaic structure gives them a sense of order and direction, but they are not working drawings. For most of his career, Repton used them to demonstrate how taste could be brought to bear on a landscape. Repton's own ideas constantly evolved, and the threads can be traced from red book to red book through the entire surviving sequence of 125 volumes. But taken individually, these red books are fragmentary—much is not discussed, and what is discussed does not always make sense.

In his published work, on the other hand, Repton invariably appealed to the good taste that he reckoned to share with his readers and clients. He was careful to smooth away any sense that he might be doing something new or contentious. As a result, little in his books reaches beyond commonly held common sense opinion. He gives the erroneous impression that landscape gardening is hardly an art and hardly requires an artist; it is just a matter of deploying an educated good taste.

Supposing instead that we prize what Repton did above what he wrote, and turn to seven major contributions to the art of which he was the master—all of them already touched on in this book (had I selected another set of fifteen landscapes to write about, the evidence might have led to another seven).

IX

The movement of water in streams and brooks, of village life, of animals across the grass, of smoke winding up in "fleecy folds" against a backdrop of trees—Repton's love of these is itself a rejection both of the timeless stillness of Brown's landscapes and of the Picturesque theories of Brown's critics, who held that taste in landscape gardening could only be acquired through the study of great painting. Instead, they signalled Repton's acceptance of the indeterminate and unfinished.

X

Repton might make random interventions to see what happened. If Brown made lakes and parks, if his houses sat close to the top of a valley side, commanding the whole sweep of the topography, then Repton might say no to the lake at Cobham, and no again to the deer park at Honing, instead placing the house perversely to obstruct the flow of the ground at Luscombe and Endsleigh, and turning it away from the view at Stanage.

XI

Brown's pleasure grounds and landscapes were worked through with a balanced harmony that gave them their timeless majesty. Repton freed himself from this harmony, writing in *Fragments* (1816) that "*Gardening* must include the two opposite characters of native wildness [in parks], and artificial comfort [in the polish of gardens]."

XII

Repton had no words to describe the effect of the "thicket of thorns, holleys and nut bushes" that he sought to introduce at Pentillie, but the surrounding landscape around the garden at Endsleigh was densely wooded, and Repton's parkland at Stanage and Brightling was to be planted over, so as to allow for natural clearings as it grew up. In this reintroduction of what might be regarded as the scenery of Britain before the intervention of man and the beginnings of farming, are the seeds of John Muir's Sierra Club, of Olmsted's plan for the conservation of Yosemite, of Niagara Falls and the Adirondacks, and of the rewilding movement in Britain today.

XIII

Repton was the first to design the English garden as it is now recognised across the United States and in England (notably in the Cotswolds): an Arts and Crafts style developed sixty years before William Morris moved to Kelmscott, with garden rooms, herbaceous borders, pergolas, and terraces that rely on vernacular materials, springing from the main house, rather than separate from it.

XIV

Repton was also responsible for that extraordinarily popular and widespread phenomenon of the front garden, such as we see at Blaise Hamlet: a polished garden between the cottage and the street, often formally planted with trimmed hedges, kept tidy and never used because the house has a side door, or a back door, by which people come and go (the front door being kept for funerals).

XV

Finally, and still more significant, Repton was responsible for popularising first the manor house and then the "natural" cottage as replacements for neoclassical and Gothic architecture. The cottage freed architecture from rule-based forms. Repton's Woodman's House at Blaise Castle, humble and vernacular, is the paradigm of a type that has been more influential in the Western world than any building since Pericles commissioned the Parthenon, and Abbé Suger the Basilica at Saint-Denis. Honest "natural" semi-detached houses (semi-detached for economy) might sometimes be ornamented with mock-Tudor woodwork in a gesture to their origin as cottages, but they never condescended to the labouring classes. Their smart front and private back gardens offered houses for a population that had grown by nearly 40%, from 7.7 to 10.7 million by the end of the Napoleonic wars, at guaranteed densities of around twelve to the acre. They occupied the leisure hours of the working family and distracted people from thoughts of revolution. These are the houses that run out in ribbons along every suburban street, and all the roads of Harrow and Watford, where I spent my childhood.

What Repton patiently fabricated out of the English countryside was a class system. He carefully distinguished every building from palace to parsonage and from country house to cottage, ensuring that these and their settings would be appropriate to their roles, having regard to the expectations and comfort of people of every class, and treating them with respect. Repton's writing and all his marketing skills have blinded us to the scale of his real achievement. Would it be too much to see him as the revolutionary, and his idea of the way we should live to have had an influence to match the vaunted triumphs of Napoleon?

— *NOTES* —

INTRODUCTION

1. John Claudius Loudon, *The landscape gardening and landscape architecture of the late Humphry Repton, esq. Being his entire works on these subjects* (London: Longman & Co, 1840), 15.

LEARNING ON THE JOB

1. *London Chronicle*, July 14, 1789, 55.
2. *The War Speeches of William Pitt the Younger*, ed. Sir Reginald Coupland, (Oxford: Clarendon Press, 1916), xvii, 16.
3. Edmund Burke, *Reflections on the Revolution in France* (London: J. Dodsley, 1790).
4. William Wordsworth, "The French Revolution as It Appeared to Enthusiasts at Its Commencement.", *The Friend* (October 26, 1809).

FERNEY HALL

1. Warwickshire County Record Office CR 1707/117 (October 7, 1806).

COBHAM HALL

1. Uvedale Price, *Essay on the Picturesque* (London: J. Robson, 1794), 57–58.
2. Richard Payne Knight, *The Landscape, a didactic poem. In three books* (London: G. Nicol, 1794) Part III, 70 [note on line 27].

EARLY MATURITY

1. Burke, 91, 113.
2. W. S. Hathaway, *The speeches of the Right Honourable William Pitt, in the House of Commons*, 3 vols. (London: Gale, 1817), vol. II, 103.
3. *The Letters of William Godwin*, vol. I: 1778–1797, ed. Pamela Clemit (Oxford: OUP, 2011), 98.
4. William Wickham, Letter to the Duke of Portland in Pelham Papers, British Library Add MS 33107, fol. 3 (January 3, 1801).

MOGGERHANGER PARK

1. Hon. John Byng, *The Torrington Diaries*, ed. Cecil Bruyn Andrewes (London: Eyre & Spottiswoode, 1938), vol. IV, 16.
2. Byng, vol. III, 11.
3. Price, 138–139.

COURTEENHALL

1. *Fragments*, 419

THE PICTURESQUE CONTROVERSY

1. John Thomas Smith, *Remarks on Rural Scenery* (London: Joseph Downes, 1797), 8–9

BLAISE

1. David Watkin, *The Life and Work of C. R.Cockerell* (London: Zwemmer, 1974), 80–81. Nash himself was Cockerell's only informant for this attribution.
2. Nigel Temple, *John Nash & the Village Picturesque* (Stroud: Alan Sutton, 1979), 115–121
3. Prince Hermann von Pückler-Muskau, *Letters of a Dead Man*, ed. and trans. Linda Parshall (Washington DC: Dunbarton Oaks, 2016),635 (December 21, 1823)
4. James Malton, *An Essay on British Cottage Architecture: being an attempt to perpetuate on Principle, that peculiar mode of building, which was originally the effect of Chance* (London: Hookham and Carpenter, 1798) 16

THE NASH YEARS

1. John Ehrman, *The Younger Pitt*, 3 vols. (London: Constable, 1969–1996), vol.2, 624.
2. Albert Sorel, *Europe and the French Revolution: The Political Traditions of the Old Regime*, ed. and trans. A. Cobban and J. W.Hunt (London: Collins, 1969), 362.

CHANGES IN TASTE

1. E. A. Smith, *Lord Grey: 1764–1845* (Oxford: Clarendon Press, 1990), 86.
2. Walpole Society vol. LXVIII (Leeds: Maney Publishing, 2006), 137 (originally published January 20, 1801).

STANAGE

1. John Adey Repton was to continue working on the house after his father's death in 1818. The house was not completed until the 1860s.
2. Knight, Book II ll, 290–293.
3. 635 "maiden oak trees fit for the navy" were auctioned from Stanage in 1796. Perhaps despite his Republican sympathies, Thomas Johnes wanted to show himself to be a patriot.

ENDSLEIGH COTTAGE

1. Jeffry Wyatt changed his name to Wyatville in 1824 and is often known by that name.

opposite: Shattered and split, this ash was old enough to be indicated by Repton on his plan for Courteenhall. It has many years of life left.

— *ACKNOWLEDGEMENTS* —

THIS BOOK WOULD have been nothing without Joe Cornish's photographs. His is a visually driven narrative to lay alongside my own text; our paths run parallel and sometimes intersect, but have never conflicted. I have placed a similar reliance on my editor at Rizzoli, Jacob Lehman, and on Robert Dalrymple's design and typesetting. The contribution of these three is so evident that perhaps it needs no acknowledgement from me. Less conspicuous, but no less important, are the volunteers and their multiform volumes of valuable research, published by England's County Gardens Trusts as part of the commemoration of the bicentenary of Repton's death in 1818.

In my own writing I have had the further encouragement of two formidable Repton scholars, Professors Steve Daniels and Tom Williamson, and I remain indebted to André Rogger, whose seminal and scholarly account of Repton's red books has brought a new rigor to the discipline of garden history.

My final debt, however, is to Gilly, at whose urging I forsook the compost heaps and traded my garden fork for a pen. JP

"Mr. Repton's Opinion of Aspects," detail from the red book for Pentillie Castle.